D1166020

A

Philip E. Lilienthal

BOOK

The Philip E. Lilienthal imprint
honors special books
in commemoration of a man whose work
at University of California Press from 1954 to 1979
was marked by dedication to young authors
and to high standards in the field of Asian Studies.
Friends, family, authors, and foundations have together
endowed the Lilienthal Fund, which enables UC Press
to publish under this imprint selected books
in a way that reflects the taste and judgment
of a great and beloved editor.

The publisher and the University of California Press Foundation gratefully acknowledge the generous support of the Philip E. Lilienthal Imprint in Asian Studies, established by a major gift from Sally Lilienthal.

Fruit from the Sands

The publisher gratefully acknowledges the generous support of the Chairman's Circle of the University of California Press, whose members are:

Elizabeth and David Birka-White
Linda Carlson Shaw
David Hayes-Bautista
Patricia and Robin Klaus
Harriett and Richard E. Gold
Susan McClatchy
Lisa See
Peter J. and Chinami S. Stern
Lynne Withey and Michael Hindus

Fruit from the Sands

The Silk Road Origins of the Foods We Eat

ROBERT N. SPENGLER III

UNIVERSITY OF CALIFORNIA PRESS

University of California Press, one of the most distinguished university presses in the United States, enriches lives around the world by advancing scholarship in the humanities, social sciences, and natural sciences. Its activities are supported by the UC Press Foundation and by philanthropic contributions from individuals and institutions. For more information, visit www.ucpress.edu.

University of California Press
Oakland, California

Library of Congress Cataloging-in-Publication Data

Names: Spengler, Robert N., III, author.
Title: Fruit from the sands : the Silk Road origins of the foods we eat / Robert N. Spengler III.
Description: Oakland, California : University of California Press, [2019] | Includes bibliographical references and index. |
Identifiers: LCCN 2018045503 (print) | LCCN 2018047116 (ebook) | ISBN 9780520972780 (ebook and ePDF) | ISBN 9780520303638 (cloth : alk. paper)
Subjects: LCSH: Food—Social aspects—History—To 1500. | Gastronomy—Social aspects—History—To 1500. | Silk Road—History—To 1500. | Agriculture—Asia, Central—History—To 1500. | Globalization—Asia, Central—History—To 1500.
Classification: LCC TX351 (ebook) | LCC TX351 .S64 2019 (print) | DDC 641.01/3—dc23
LC record available at https://lccn.loc.gov/2018045503

Manufactured in the United States of America

26 25 24 23 22 21 20
10 9 8 7 6 5 4 3 2

CONTENTS

A WORD ON SEMANTICS

A general shortcoming of the social sciences (as opposed to the physical and biological sciences) is a lack of widely agreed-upon nomenclature. The meaning of one term to an individual archaeologist or social theorist may not be shared by others. To forestall any misconceptions, I open with a brief list and definition of problematic terms in the field of Central Asian archaeology.

CENTRAL ASIA Politically defined geographic region encompassing the former Soviet republics of Kazakhstan, Kyrgyzstan, Tajikistan, Turkmenistan, and Uzbekistan.

CENTRAL EURASIA An inconsistently used geographic term encompassing Central Asia as well as adjacent regions to the north and the east, notably western Mongolia, the Tuva region of Russia, and Xinjiang, Qinghai, and Tibet in China.

EXCHANGE The transfer of goods, ideas, genes, or cultural traits between different people, whether traded, sold, coerced, bartered, stolen, or gifted.

INNER ASIA A geographic term often employed by historians and occasionally by other scholars. The term loosely parallels *Central Eurasia*.

NEW WORLD The Americas or the Western Hemisphere of the globe—the parts of the world generally unknown to people in Europe or Asia before the expeditions of Christopher Columbus.

OLD WORLD A term usually referring to Europe, Asia, and Northern Africa.

SILK ROAD In the traditional sense, an ancient network of exchange routes connecting China to the Mediterranean; historians often claim that these originated with the Han Dynasty of China in the second century BC. In this volume, I use the term more broadly to encompass exchange through Central Asia from the third millennium BC until the modern era.

SPICE ROUTES An arbitrary term designating exchange routes parallel to but south of the Silk Road, by which a wide variety of plants were brought from across South Asia to Europe. Although exchange peaked later along the Spice Routes than along the Silk Road routes, it was essentially the same social process.

A NOTE ON DATES

While terms such as *Bronze Age* and *Iron Age* are often used to denote phases in European or West Asian history, they are not used consistently in different regions or even among scholars. Hence, I generally do not use them in this book; instead, I specify millennia or exact dates.

I have also opted not to use terms denoting Central Asian archaeological culture groups, such as Srubnaya and Andronovo, in the text; however, I use terms denoting historically documented dynasties and periods in classical antiquity.

PERSIAN EMPIRES

Median (728–549 BC)

Achaemenid (550–330 BC)

Parthian (247 BC–AD 224)

Sasanian (AD 224–651)

Samanid (AD 819–999)

CALIPHATES

Rashidun (AD 632–61)

Umayyad (AD 661–750)

Abbasid (AD 750–1258)

Ottoman (AD 1517–1924)

CHINESE DYNASTIES

Xia (ca. 2000–1600 BC)

Shang (ca. 1600–1046 BC)

Zhou (1046–256 BC)

Warring States period (475–221 BC)

Qin (221–206 BC)

Han (206 BC–AD 220)

Three Kingdoms period (AD 220–80)

Jin (AD 265–420)

Southern and Northern Dynasties (AD 420–589)

Sui (AD 581–618)

Tang (AD 618–907)

Five Dynasties period (AD 907–60)

Song (AD 960–1279)

Yuan (AD 1271–1368)

PERIODS IN CLASSICAL ANTIQUITY

Classical Greece (410–323 BC)

Hellenistic period (323–146 BC)

Roman Republic (509–27 BC)

Roman Empire (27 BC–AD 476)

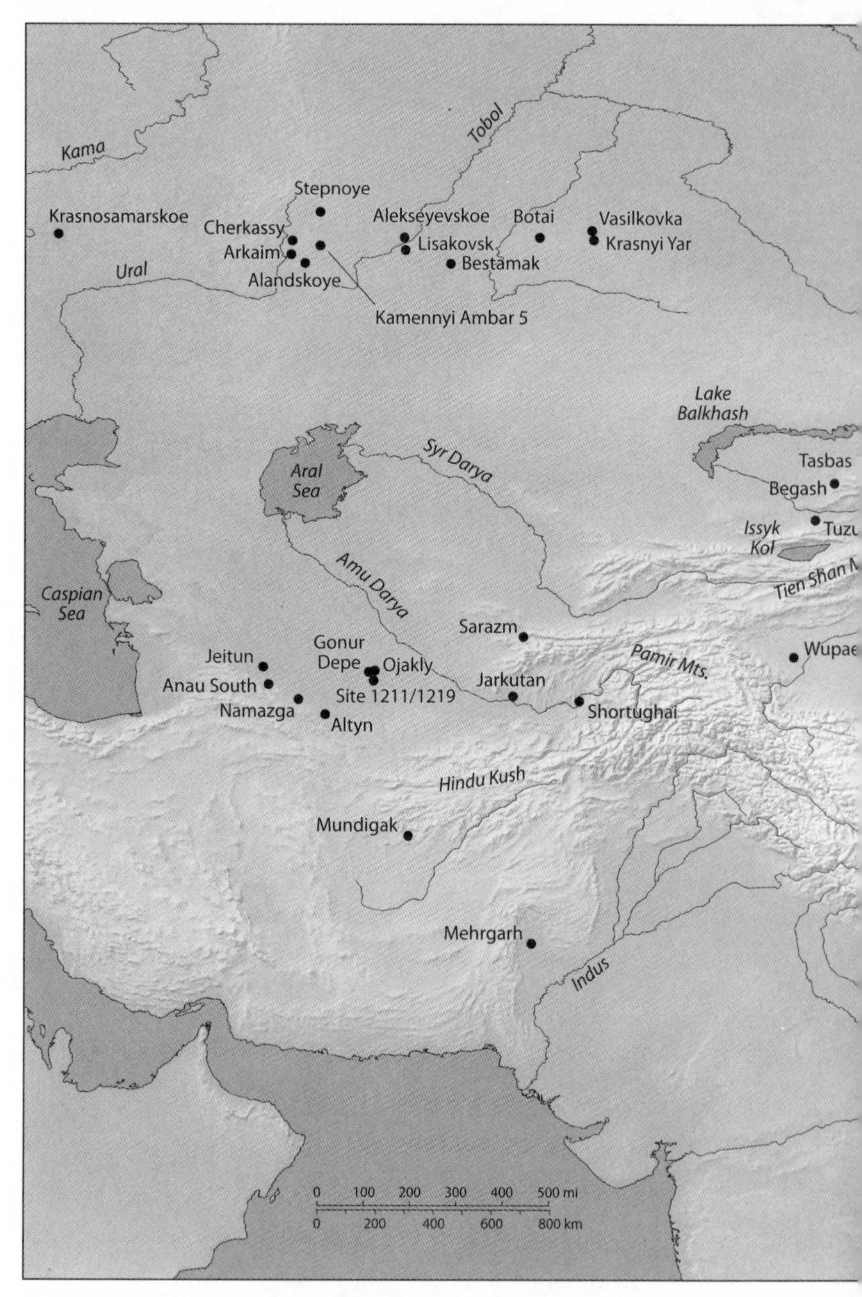

Map 1. Map of Central Eurasia, showing key archaeological sites and geographic features. These sites have provided the archaeobotanical evidence I have used to trace the spread of agricultural crops across Inner Asia.

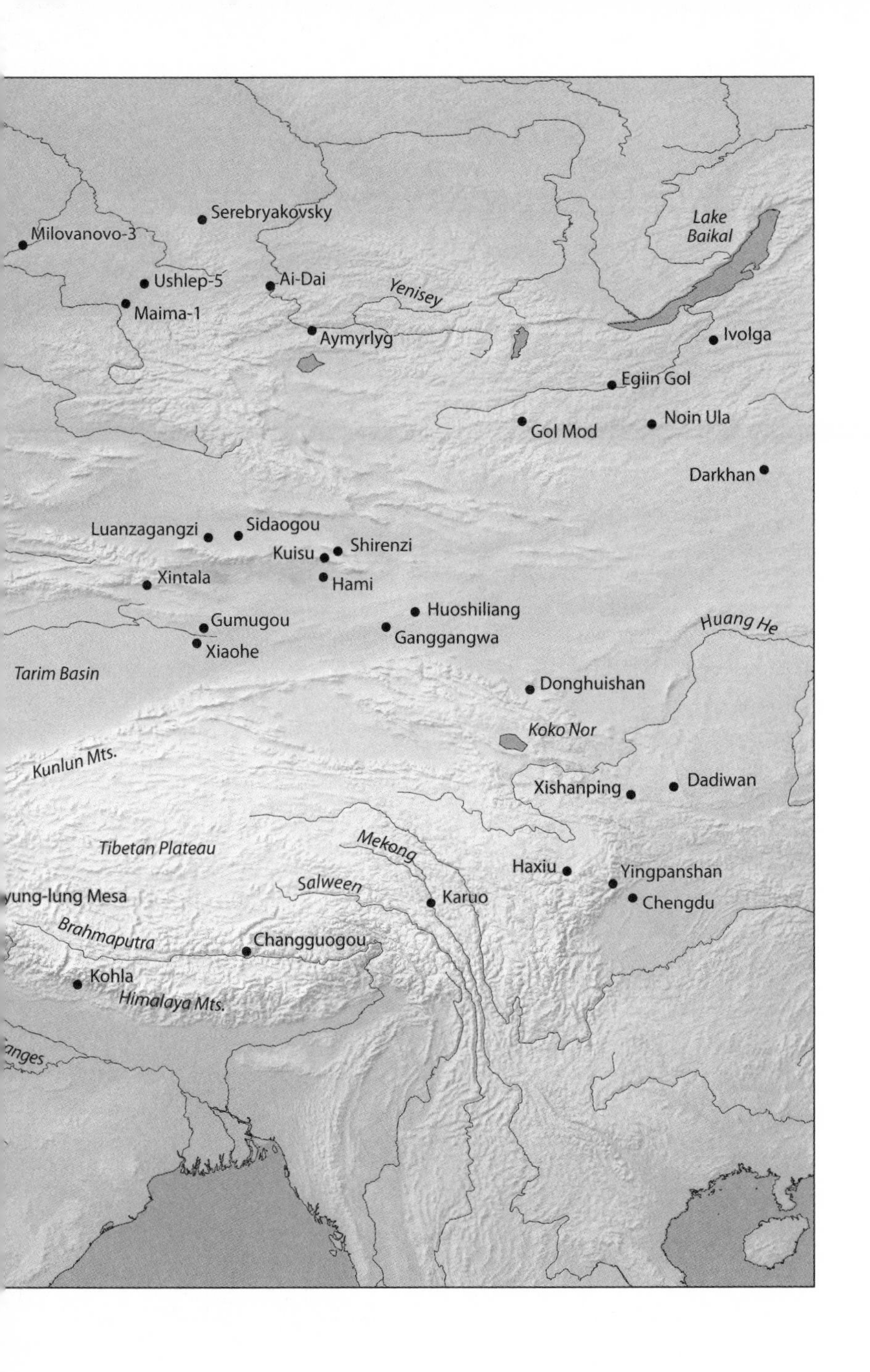

Serebryakovsky
Milovanovo-3
Ushlep-5
Ai-Dai
Maima-1
Yenisey
Aymyrlyg
Lake
Baikal
Ivolga
Egiin Gol
Gol Mod
Noin Ula
Darkhan
Luanzagangzi
Sidaogou
Kuisu
Shirenzi
Xintala
Hami
Huoshiliang
Gumugou
Ganggangwa
Xiaohe
Huang He
Tarim Basin
Donghuishan
Koko Nor
Kunlun Mts.
Xishanping
Dadiwan
Tibetan Plateau
Mekong
Haxiu
Yingpanshan
Salween
Karuo
Chengdu
Brahmaputra
Changguogou
Kohla
Himalaya Mts.

PART I

How the Silk Road Influenced the Food You Eat

CHAPTER ONE

Introduction

We have all heard the rhetoric about sustainable food systems and the slogans "Feed the World" and "Nine billion by 2050." Likewise, we have heard about the "loss" of cultures around the world through the domination of one globalized culture that shapes every aspect of our daily lives, including how and what we eat. As the human population approaches nine billion, we are seeing unprecedented rates of deforestation in South America for the sole purpose of planting fields of soybean (*Glycine max*), an East Asian domesticate. Genetic diversity is being lost among crops around the globe as cloned and genetically identical hybrids are planted in fields from Russia to Mexico, and fruits and vegetables that were unheard-of in the northern temperate zone a generation ago are available all year round in markets and grocery stores. But how did humanity get to this point? How have we reached this fever pitch of global communication, commerce, and resource distribution? How did humans gain the ability to reshape the ecosystems around them and even the climate of the earth itself? The answers to these questions lie buried in the archaeological record.

Cherry-picking concepts from the literature of world systems, we can think of the globe today as one system or mechanical circuit. Within this system, poor economic decisions by a political leader in America or risky gambles by corporate bankers can send the entire globe into an economic crisis. A drought in California directly affects the price of oranges in

New York City. America's addiction to caffeine has caused the burning of millions of acres of ancient rainforests from Hawaii to Brazil. The combustion of fossil fuels in the United States and Western Europe is directly responsible for the melting of glaciers in the Himalaya and the decimation of tropical coral reefs, and it is in the process of submerging entire island nations.

Focusing on food systems, I look at globalization in Asian prehistory and history, starting about 4,500 years ago. When broken down to their simplest components, the ancient and modern processes of culture and gene flow are similar. Exchange along the Central Asian trade routes that came to be known as the Silk Road expresses, in its most basic sense, what Karl Marx referred to as commodity fetishism, the desire to obtain the exotic. People are willing to pay exorbitant sums to procure goods that their neighbors do not have. Commodity fetishism and conspicuous consumption are as much a part of the modern world as they were of the ancient world (if not far more so), and every reader of this book participates in them, although to varying degrees. Many scholars have studied how the exchange networks of the ancient world operated and connected people. But there remains the theoretical question of when the world system of Asia formed: when southwest, South, and East Asia became interconnected.

During the Yuan Dynasty (1271–1368)—the period of Mongol reign in China—Eurasia fell under one economic system, and the Mongolian trade routes led to all corners of Asia. By 1280, China, Central Asia, and Iran were under Mongol control—the *Pax Mongolica*—and trade routes stretched into more distant regions. Genghis Khan conquered the vast majority of Asia, occupying northern China by 1234, and his grandson assumed the Chinese throne by 1279. After the Mongol rulers (the Ilkhanate) pushed out the last of the Abbasid Caliphate from southwest Asia, the Asian continent was unified for the only time in history.

Yet Asia was deeply interconnected long before the formation of the Mongol khanates. In AD 751 the clash between the Arab and Chinese (Tang) Empires at the Talas River in Central Asia resulted in the full Islamization of Central Asia. This was the only military encounter between the great powers of the ancient world, but these empires and their predecessors had been economically connected for at least a millennium: trade along the Silk Road predated the Han Dynasty. Historical accounts claim that Zhang Qian, the first political envoy from the Chinese center, reached Central Asia in 126 BC. The written accounts in the *Shiji* (The records of the Great Historian, ca. 80 BC), although embellished with descriptions of fantastic beasts and other mythical details, attest to direct communication between Central Asia and East Asia in the early Han period. While these accounts may suggest that Central Asia was previously unknown to the Han, the mountains of Central Asia never impeded culture flow: people had been moving through the valleys like water through a leaky faucet for at least two millennia before Zhang Qian's journey to the west.

Archaeological evidence shows that the food systems of these regions were influencing each other as far back as the early third millennium BC. Around 2200 BC, at a small settlement in the Dzungar Mountains of northern Kazakhstan (at a site named Begash by archaeologists in the early 2000s), a family of farmers and herders ate bread made from grains that likely grew in a nearby field: broomcorn millet, a crop that was domesticated several millennia earlier in northeastern China, and wheat, a crop originating in the Fertile Crescent of southwest Asia.[1] Finding these two grains together provided the earliest evidence of a food system extending across a continent. The Silk Road remained a conduit for the mingling of the cuisines of different regions of East and Central Asia until the fourteenth century. Broomcorn millet continued to travel west through this corridor and eventually became a major crop in the Roman Empire and across Europe. In the opposite direction, wheat was brought to East Asia and transformed into noodles and dumplings, dramatically influencing Chinese cuisine.

Increased connections between the food systems in Eurasia and beyond led to further globalization of regional cuisines. Eventually, rice (*Oryza sativa*) would become one of the most important grains in the Islamic world; the apple (*Malus domestica/pumila*) would become a symbol of America; and the peach (*Prunus persica*) would become an emblem of the US state of Georgia, while its cousin, the apricot (*Prunus armeniaca*), would become a hallmark of the nation of Georgia in the Caucasus. How each of these crops migrated from its place of origin in East or Central Asia is a fascinating story, requiring both an archaeological and historical perspective. In this book, I trace the origins and the paths of dispersal of these and other familiar crops. I illuminate how the farmers of the ancient world developed the foods we eat today and the role the Silk Road played in their evolution and dispersal.

THE BEGINNINGS OF THE SILK ROAD

> When man migrates, he carries with him not only his birds, quadrupeds, insects, vegetables, and his very sward, but his orchard also.
>
> Henry David Thoreau, "Wild Apples," 1862

By dispersing plants and animals all around the world, humans have shaped global cuisines and agricultural practices. One of the most fascinating and least-discussed episodes in this process took place along the Silk Road. This story has come to light through recent discoveries in the fields of archaeology and biology—notably in the areas of phytogenetics and archaeobotany. I track a selection of plants on their historical journey along the trans-Eurasian exchange routes, revealing how the food on your kitchen table made its way across deserts, over mountains, and through thousands of farming seasons, and how the introduction of new crops changed the course of human history.

In 2001, Michael Pollan's book *The Botany of Desire* enabled readers all over the world to learn how the apple made its way into our kitchens, and, according to Pollan, helped settle the American frontier.[2] Many readers were surprised to learn that the story of the apple tree reaches back to Central Asia, and that modern commercial apples have genetic linkages to truly wild populations outside Almaty, the former capital of Kazakhstan. In fact, it is the Silk Road itself that is responsible for the genesis of our modern apple, which is a hybrid of four separate populations. When the Silk Road merchants carried apple seeds across Eurasia, the trees hybridized with populations that had been isolated since the last glaciation of Eurasia, creating offspring that produced more and larger fruits.

Your grandmother's apple pie is not the only food on your table to trace its roots back to Central Asia, nor the only one to travel the great Silk Road. Pistachios (*Pistacia vera*) originated in the foothills of southern Central Asia, and almonds (*Prunus dulcis*) and English walnuts (*Juglans regia*) trace their lineages back to the foothills of southern Eurasia.

The Silk Road was the largest commerce network of the ancient world; it linked the edges of the Eurasian supercontinent to trading centers in Central Asia and indirectly connected the imperial centers of East and southwest Asia. While organized trade along the Silk Road, along with military outposts and government taxation, dates to the Han Dynasty (206 BC–AD 220), archaeologists have traced the dispersal of goods, ideas, cultural practices, and genes across Central Asia to the third millennium BC.[3] I treat these earlier iterations of the Silk Road as precursors of the historical routes, and I give cultural exchange during the second and third millennia equal prominence with that of later periods. Over the past two millennia, control over the lands of the Silk Road, with their expansive deserts and mountain ranges, has passed back and forth between various political and ethnic groups, including successive East Asian empires and Central Asian political entities, such as the Xianbei and Xiongnu. This ebb and flow of cultures shaped human history in myriad ways, including the spread of agricultural practices and crop varieties.

The Silk Road was not a road, nor was its main commodity silk. The popular image of long ribbons of camel caravans connecting Xi'an with Rome is just one of its temporal iterations. I define the Silk Road loosely, as a cultural phenomenon of exchange and interaction starting in the third millennium BC and intensifying during the first millennium BC, as exchange and mobility (in various forms) turned Central Eurasia into a complex social arena. The gradual increase in human mobility across Inner Asia, initially propelled by the spread of horse transport, seasonal human migration, and small-scale agropastoralism, played a significant role in shaping human history. Prehistoric Central Asians connected the far corners of the ancient world and spread innovations across it. Among the many ideas and technologies they picked up was the knowledge of how to grow and experiment with crops. Many of these crops were later transported and introduced to new geographic areas.

The chapters that follow trace the seven-thousand-kilometer journey that humans have been embarking upon for millennia (although few merchants likely traveled the entire route). They track the footsteps of European explorers like Marco Polo, Alexander von Humboldt, Sven Hedin, Aurel Stein, Nikolay Przhevalsky, Nikolai Ivanovich Vavilov, Owen Lattimore, and hundreds of thousands of merchants and herders who carried with them the genetic material for new varieties of plant and animal species. The resulting dispersal of organisms by humans was unmatched until the colonial expansion of Europe. Notably, the grain crops they carried would give rise to crop-rotation cycles, increasing the food supply and allowing empires to flourish in both Europe and Asia. Millet became the summer crop of the Persian Empire and the low-class crop of Rome, and wheat became the

Figure 1. A funerary figurine of a Sogdian Silk Road merchant on the back of his Bactrian camel, from the Tang Dynasty (AD 618–907). The ceramic figure is colored with a characteristic three-color (*sancai*) lead glaze. Art Institute of Chicago/Art Resource, NY.

winter crop of the Han and later dynasties in China. Seeds were kept warm through harsh winters at high elevations and irrigated in some of the driest sands of Asia.

Through their thirst for knowledge and their desire to adapt to some of the most rugged landscapes on the globe, prehistoric Central Asians spread the allele for a highly compact morphotype of wheat (*Triticum aestivum*). They experimented with drought-tolerant millets from East Asia and brought the first peach tree to southwest Asia. Peaches that originated along the marshes of the Yangtze River Valley in Zhejiang, China,[4] made their way to Europe by the Hellenistic period. Like the ambrosia of classical mythology, the peach was a symbol of immortality in Daoism, and the flesh of the peach was immortalized in ink in the *Shijing* (The book of odes), written in the first millennium BC.[5]

The peach is not the only fruit in today's Euro-American cornucopia to have followed the great Silk Road. According to the Han historian Sima Qian, the military ambassador Zhang Qian, returning from one of his diplomatic missions to Central Asia in the later second century BC, carried a short, tendrilling vine in a rawhide sack, protecting it from the hot sun as he traversed the deserts. According to Sima Qian, grapes, and the sweet wine they yielded, came to China from the country of Dawan, which most historians agree to be the Fergana Valley in modern Uzbekistan.[6] Evidence from a recently excavated grave from the Yanghai cemetery in Xinjiang suggests, however, that wine was revered in the oases of the Taklimakan Desert hundreds of years before Zhang Qian's largely mythical journey to the west.[7]

An understanding of the origins of our food links us more closely with our past and with the farmers who planted seeds and selected plants for hardiness or for sweeter, larger, or faster-growing fruits. Across Eurasia, for over ten thousand planting seasons, people have sown seeds, nurtured seedlings, decided when to plant and what seed lines to save for the following year, and passed their accumulated knowledge and improved crop varieties down to their children. A peach cobbler or a glass of wine is an archaeological artifact, containing in the genes of the fruit a narrative beginning deep in the past. Scholars around the world are striving to read that narrative to learn how humanity and our modern plant varieties evolved together.[8]

Intro:

- Archaological Evidence = shows how "food systems" influenced one another;
- The Silk Road = a network for this → my question
- How were things grown in new places?
- Grains Traveled to Europe, Asia = Cultural influences

Culture, trade, and food = intertwined

CHAPTER TWO

Plants on the Silk Road

Central Asia is a land that few students have ever learned about in school. This massive region not only contains some of the earth's most breathtaking vistas, but it also holds volumes of human history. The rocky crags witnessed the treks of countless camel caravans and generations of herders moving their donkeys, camels, cattle, goats, horses, sheep, and yaks up and down the mountains with each passing season. The desert oases of Central Asia gave rise to the legendary Silk Road cities of Bukhara, Khiva, Loulan, and Samarkand; many of these cities moved with the sands, like the wandering lake of Lop Nor described by the Swedish explorer Sven Hedin. The Gobi, Kara-Kum, Kyzyl Kum, and Taklimakan Deserts contained shifting sands, mirages, and isolated sanctuaries of vineyards and orchards. These blowing sands have buried great empires: they witnessed the rise and fall of the Greco-Bactrian state, the kingdom of Timur, and successive Persian dynasties. They stopped the advance of Alexander the Great and set the backdrop for the journey of Marco Polo and his father. More recently, they have been the front lines of the many proxy wars between the British and Russian Empires during the Great Game in the nineteenth century and the imperialist American and Soviet powers during the Cold War.

Despite the seeming desolation of expanses of Central Eurasia today, parts of it were a veritable Garden of Eden for millennia. Until the first millennium BC, much of southern Central Asia was a lush expanse of short, shrubby forests, which included wild pistachio, almond, cherry (*Prunus* spp.), and English walnut trees. Largely because of intensive human economic pursuits, the region today is mostly populated by lizards, snakes, and desert saxaul trees (*Haloxylon*). The piedmont of Central Asia was once covered in forests of sea buckthorn (*Hippophae rhamnoides*), Russian olives (*Elaeagnus angustifolia*), wild apples (*Malus* spp.), hawthorn (*Crataegus* spp.), mountain ash (*Sorbus* spp.), and a wide variety of nut-bearing trees.[1] While these forests are all but gone today, rich pockets of agricultural land continue to produce fruits from their descendant species, as well as grapes, pomegranates (*Punica granatum*), and a wide variety of sweet melons (*Cucumis melo*).

The *Memoirs of Babur*, or *Baburnama*, compiled between 1483 and 1530, chronicles the travels of Zahiruddin Muhammad Babur, who observed: "Grapes, melons, apples, and pomegranates, all fruits, indeed, are good in Samarkand; two are famous, its apple and its ṣāhibī (grape). Its winter is mightily cold; snow falls but not so much as in Kābul; in the heat its climate is good but not so good as Kābul's."[2] Samarkand, the capital city of the vast empire built by Timur (Tamerlane, 1320/1330–1405), was set on a rich oasis watered by the Zerafshan River. For centuries, the city has shone bright against the desert sands. In Babur's time it was a center of education and commerce. In the heart of the city, Timur and his successors constructed the Registan, an Islamic university as ornate as any of the palaces built in Europe at that time. Ruy González de Clavijo, a Castilian ambassador who visited Timur at his court from 1403 to 1405, wrote in his account of the journey, *Embassy to Tamerlane*, that the city of Samarkand was a bustling metropolis with beautiful gardens inside and outside the city, and orchards all around.[3]

Babur praised the wondrous variety of fruits and nuts grown across Central Asia in the fifteenth and sixteenth centuries. He noted the sweetness of the varieties of melons and specific varieties of apples. When discussing Kabul in central Afghanistan, he states that "the fruits of the cold districts in Kābul are grapes, pomegranates, apricots, peaches, pears, apples, quinces, jujubes, damsons, almonds, and walnuts; all of which are found in great abundance."[4] Babur's (1487–1530) great-grandson, the Mughal emperor Nuruddin Muhammad Jahangir (1569–1627), also praised the flavors of Central Asia. In his autobiography, while laying out the political history of the time, he discusses the cultivation of exceptionally sweet apricots, peaches, melons, and apples from the Samarkand region, as well as rice, millet, and wheat. Recalling a social encounter, he states that "on one tray they brought many kinds of fruit—Kārīz melons, melons from Badakhshan and Kabul, grapes from Samarkand and Badakhshan, apples from Samarkand, Kashmir, Kabul, and from Jalalabad, which is a dependency of Kabul, and pineapples, a fruit that comes from the European ports." Jahangir praises the apples in Kabul as second to those from Samarkand, of which he states, "I had until now never seen such delicate and delicious apples. They say that in Upper Bangash, near Lashkar-dara, there is a village called Sīv Rām, in which there are three trees of this apple, and although they have made many trials, they have never found so good ones in any other place."[5]

González de Clavijo further noted that the ornate Persianate gardens and orchards were irrigated by a complex system of conduits. One of the most significant achievements of the Qarakhanid period, preceding the arrival of Timur, was the development of elaborate irrigation systems and the expansion of farming into the deserts. The best-known example is a one-hundred-kilometer canal, dug in the Taraz region of southern Kazakhstan, but others include major extensions to the existing network of canals dug through the Fergana lowlands.[6] Abu Bakr Muhammad ibn Jafar Narshakhi's *History of Bukhara*, written ca. AD 940, describes the lavish residence of the Sassanid elites and the opulent lifestyle in the central citadel. It also notes that there were a thousand shops in the city, with vegetable vendors near the city walls, pistachio dealers not far from them, and spice merchants in their own area. The entire city was divided into districts separated by walls.[7]

Agricultural practices after the Mongolian conquests in Central Asia and Iran can be pieced together from a handful of preserved Persian agricultural manuals, such as the *Irshad al-Zira'a* (Guide to agriculture), composed in 1515 in Herat, Afghanistan, by Qasim b. Yusuf Abu Nasri Harawi. The book talks about irrigated gardens and elaborate pavilions, features of this landscape that have been lost to the political turmoil of the nineteenth and twentieth centuries. He discusses the cultivation of wheat, barley, millet, rice, lentils, and chickpeas; viticulture merits its own section. He also talks about garden crops, including cucumbers (*Cucumis sativus*), lettuce, spinach (*Spinacia oleracea*), radishes (*Raphanus raphanistrum*), onions, garlic, beets (*Beta vulgaris*), and eggplant (*Solanum melongena*); herbs and aromatic plants; hemp (*Cannabis sativa*); alfalfa (*Medicago sativa*); dye plants such as madder (*Rubi* sp.), indigo (*Indigofera* sp.), and henna (*Lawsonia inermis*); and fruits and nuts, including melons, pomegranates, quinces (*Cydonia oblonga*), pears (*Pyrus* sp.), apples, peaches, apricots, plums, cherries, figs (*Ficus* spp.), mulberries (*Morus* sp.), and pistachios.[8]

Under the tenth-century Qarakhanid ruler Shams al-Mulk Nasr b. Ibrahim, the gardens of Samarkand were expanded, and a large-scale game and hunting preserve was created. In fact, Suzani Samarkandi (Shams al-Din Muhammad, d. 1166), the only poet of the Qarakhanid court whose writings survive today, stated that Samarkand was "a paradise on Earth."[9] These Persian-inspired gardens saw their peak during the Timurid period, after a period of abandonment during the Mongol conquests. The fourteenth-century gardens in Samarkand and Bukhara were boxed-in plots of irrigated land in densely populated districts of the city.[10] Historians claim that the Timurid legacy of elaborate gardens harkens back to the much earlier forms of the Achaemenid and Sasanian periods; many of these gardens had raised flower beds that were divided into four rectangular quadrants, with irrigation canals down the middle, and wooden or stone walkways.

Today Samarkand is a city of over 350,000 people in eastern Uzbekistan; the central market houses rows of merchant vendors selling the same fruits that Babur praised nearly half a millennium ago. Its vendors still boast of the quality of the melons; they offer some of the juiciest pomegranates in Asia and a unique golden and succulent variety of peaches. When fresh fruit is out of season, the merchants sell raisins, prunes, dried apricots, apples, figs, dates (*Phoenix dactylifera*), walnuts, pistachios, almonds, and a vivid array of colorful legumes and grains (figure 2). Orchards growing the same varieties of fruits would have been established by the time Alexander the Great conquered the city, at that time called Marakanda, in 329 BC.

Samarkand was not the only ancient oasis city to have orchards and vineyards; they were an important part of all the urban centers and small towns across Central Asia. These trading posts, which served as nodes connecting the ancient trade routes, were known for distinctive local varieties of fruits, some of which were praised across the Old World. Remnants of these ancient town orchards still exist: in 1900, the Silk Road explorer and archaeologist Aurel Stein trekked through the Taklimakan Desert and stopped at the ancient Buddhist town and trading center of Dandan Oilik, also visited by Sven Hedin in 1896. Stein excavated coins that dated the site to between AD 713 and 741. He discovered that the ancient orchard trees were still standing in rows, half buried in sand. He suggested that the rows of thousand-year-old trees looked like peaches, plums, apricots, and mulberries. Stein, however, was excited by this observation largely because it promised him a steady supply of firewood for the cold desert nights.[11]

In the early 1600s Fredrick, the Duke of Holstein (in present-day northern Germany), sent a series of ambassadors to meet with the Persian rulers of the time. One of these envoys, Adam Olearius, stated that "pomegranate-trees, almond-trees, and fig-trees grow there without any ordering of cultivation, especially in the Province of Kilan [now Tehran, Iran], where you have whole

Figure 2. Dried fruit and nut vendors at the central bazaar in Bishkek, Kyrgyzstan, during a cold winter day in 2015. Photo by Julia McLean.

forests of them. The wild pomegranates, which you find almost everywhere, especially at Karabag, are sharp or sowrith."[12] Even after centuries of turmoil, the remnants of orchards and home gardens can still be seen across Turkmenistan, Afghanistan, Iran, and Iraq, in nearly every river valley and adjacent to every spring.

Discussing the market bazaars of Kashmir in 1597, Abul Fazl ibn Mubarak, Allámi, the grand vizier of the Mughal emperor Akbar, clearly describes the thriving trade in fruits, noting that grapes, melons, pomegranates, apples, pears, quinces, peaches, and apricots were carried from modern-day Uzbekistan and central India to markets in Kashmir (in the northwestern Indian subcontinent). His description attests to the merchants' ability to transport perishable goods over great distances, even in the desert heat.

> Melons and grapes have become very plentiful and excellent; and watermelons, peaches, almonds, pistachios, pomegranates, etc., are everywhere to be found. Ever since the conquest of Kábul, Qandahár, and Kashmír, loads

> of fruits are imported; throughout the whole year the stores of the dealers are full, and the bázárs well supplied. Muskmelons come in season, in Hindústán, in the month of Farwardín (February–March), and are plenty in Urdíbihisht (March–April). They are delicious, tender, opening, sweet smelling, especially the kinds called náshpátí, bábáshaikhí, 'alíshérí, alchah, barg i nai, dúd i chirágh, etc. They continue in season for two months longer. In the beginning of Sharíwar, (August) they come from Kashmír, and before they are out of season, plenty are brought from Kábul; during the month of A'zar (November) they are imported by the caravans from Badakhshán, and continue to be had during Dai (December.) When they are in season in Zábulistán, good ones are also obtainable in the Panjáb; and in Bhakkar and its vicinity they are plentiful in season, except during the forty cold days of winter. Various kinds of grapes are here to be had from Khurdád (May) to Amurdád (July), whilst the markets are stocked with Kashmír grapes during Shahríwar. Eight sérs of grapes sell in Kashmír at one dám, and the cost of the transport is two rupees per man. The Kashmírians bring them on their backs in conical baskets, which look very curious. From Mihr (September) till Urdíbihisht grapes come from Kábul, together with cherries, which his Majesty calls sháhálú, seedless pomegranates, apples, pears, quinces, guavas [this is probably a translation error and may refer to a citrus], peaches, apricots, girdálús, and álúchas, many of which fruits grow also in Hindústán. From Samarqand even they bring melons, pears, and apples.[13]

Although historical sources attest to the diversity of items moving along the medieval Silk Road, they provide little evidence of what actual goods were moving along the earlier trade routes. Of the texts surviving from the late first millennium AD, one of the most informative on the trade along the Silk Road is a pamphlet titled *The Investigation of Commerce* (*al-Tabassar bi'l-Tijara*). The scholar al-Tabari (AD 839–923) attributes the text to the Arab writer al-Jahiz (AD 776–868/869), who wrote extensively on biology, theology, and philosophy. He produced over two hundred books, mostly during his fifty years in Baghdad, at the heart of the Abbasid Caliphate.[14] Charles Pellat, who translated the text into French in 1954, acknowledges that its authorship is questionable but notes that even if al-Jahiz did not write it, the text does date to the ninth century. He suggests, however, that al-Jahiz was not personally familiar with the commercial trade routes of the ancient world. Most likely, he talked with traveling merchants about the items brought into Baghdad from across Asia, particularly luxury goods.

Among the goods discussed are pearls originating from several different South Asian bodies of water (whose value depended on their origins), as well as gemstones such as carnelian, topaz, turquoises of varying shades, garnets, diamonds, and various crystals. Al-Jahiz also mentions amber, gold, and musk from Tibet. He discusses the transport of silk garments; furs and pelts, including rabbit, black fox from the Caspian Sea region, ermine, and panther; and tapestries from Armenia. These textiles, varying widely in color and design, incorporated fibers from all across Inner Asia. They included Byzantine tapestries with purple stripes on a red and green background, felted fabrics from western China, and silk from Isfahan, Iran.

Among the many other items mentioned are balm of Gilead from the Arabian balsam tree (*Commiphora gileadensis*) and processed sugar, which had likely reached this region from India about a millennium earlier but was probably for elite consumption only. Other imports from India mentioned in the text include tigers, elephants, and panthers, sandalwood (*Santalum* sp.), ebony (*Diospyros* sp.), and coconuts (*Cocos nucifera*). From China came spices, notably cinnamon (*Cinnamomum* sp.), along with silks, ceramics, paper, peacocks, horses, saddles, felt, and rhubarb (*Rheum rhabarbarum*); from Byzantium, gold and silver utensils, coins, and ornaments, along with red copper, lyres, female slaves, craftsmen, and eunuchs; from Arabia, horses, ostriches, skins, and wood; Arabic merchants obtained panther skins, hawk feathers, and felts from the Berbers. From Yemen came giraffes, coats, skins, carnelian, incense, indigo, and turmeric (*Curcuma longa*), and from Egypt came papyrus (*Cyperus papyrus*), topaz, and balsam. From the Khazar Khaganate in southern Central Asia the Abbasid elites obtained slaves, armor, mesh (and likely other metal goods), fennel (*Foeniculum vulgare*), and sugarcane, suggesting that sugar was already being grown and processed in the region of modern Turkmenistan by the ninth century AD. From Samarkand came paper; from Balkh in Afghanistan, grapes and mushrooms; from Merv, fabrics (likely cotton) and silk, pheasant, and jujube fruits (*Ziziphus* sp.); and from Khorasan in northeastern Iran, wool coats. Goods from lands in modern-day Iran included honey, quinces, apples, Chinese pears (*Pyrus* sp.), salt, saffron (*Crocus* sp.), fruit syrups, and clothing from Isfahan; indigo, cumin (*Cuminum cyminum*), dried and fresh fruits, rosewater, linen, jasmine oil syrups, glassware, silks, and sugar from Kerman; cedar wood, violet oil, and horse blankets from the ancient city of Susa; and sugar, dates, grapes, and dancers from Ahvaz.

Around this time a culinary revolution took place, likely centered in Baghdad. It was sparked in part by the introduction of new crops, like sugarcane and rice, but also by new irrigation systems that made it possible to grow them. The Abbasid caliphs imported expert cooks from all corners of their empire and refined the art of Arabic cooking. The oldest surviving Arabic cookbook, the *Kitab al-Tabikh* (The book of dishes), was written by Muhammad bin Hasan al-Baghdadi in 1226.

Much of our understanding of life in southern Central Asia and the Iranian Plateau during the first millennium AD comes from the writings of Islamic geographers.[15] One of these, al-Muqaddasi (AD 945–91), who was employed by the court of Adud al-Dawla (936–83) in Shiraz, wrote about the construction of irrigation systems, caravansaries, and dams in the lower Kur River in the Fars region of Iran around 985. He wrote about the ancient city of Istakhr in Iran, where irrigation systems flowed around the city, watering rice fields and orchards.[16] The Pulvar River, a tributary of the Kur, runs in a bend around the ruins of Isatkhr today. The name means "pool" in Farsi, implying that there might once have been major water catchments near the city.

As elaborate irrigation technology spread farther into Central Asia, along with Islam, cuisines changed. The prominence of rice in Central and southwest Asian dishes likely dates to this period. A 1972 survey of ninth-century Islamic agriculture in Iraq by Husam Qawam El-Samarraie draws primarily on Ibn Wahshiyya's early-tenth-century *Book of Nabatean Agriculture* (*Kitab al-filaha al-Nabatiyya*).[17] Ibn al-Awam, an Iberian author, compiled an agronomic treatise, *Kitab al-Filaha* (Ibn al-Awam's book of agriculture), based on information gathered from ancient and contemporary authors at the end of the twelfth century or the first half of the thirteenth century.[18] Abu Hamid al-Andalusi al-Gharnati was born in 1080 and started traveling in 1106; from 1130 to 1155 he was in Khorasan, where he noted that there were many cities, villages, farmsteads, and fortresses. He stated that there were "fruits, the like of which I have not seen in any of the other countries I have visited." He noted watermelons sweeter than "sugar with honey," large melons that would keep all winter, large dates, red and white grapes, apples, pears, and pomegranates.[19]

About a century earlier, in 988, another Arabic geographer, Ibn Hawqal (who traveled from 943 to 969) had described Khwarazm as a "fertile country, producing many kinds of grain and fruit, but not walnuts. Its cotton and woollen textiles are exported throughout the world."[20] Ibn Hawqal was one of the few authors of the time who traveled near the Middle Euphrates River. He recorded agricultural details in his *Arrangement of the World* (*Kitab surat al-ard*).[21] Ibn Battuta, another of these ancient geographers, traveled around South Asia and North Africa from 1325 to 1354 and praised the qualities of the fruits across the Islamic world. His journeys never took him into the heart of Inner Asia, however, so we cannot know whether he would have compared these fruits favorably with those of the regions to the north.

Along with these geographic writings, a number of Arabic cookbooks were written and survive in part or in full today. Most of these books were produced in the thirteenth century. Catering to the elite tastes of the caliphate, they draw on ingredients available on the Iranian Plateau and west across to the Mediterranean coast. From these works we can extrapolate some details about what the common people in Central Asia would have consumed. Two of the most important volumes were produced in the Mediterranean and three in the eastern Islamic world.[22] One of the most detailed, *Kitab al-Wuslah ila l-Habib fi Wasf al-Tayyibat wal-Tib*, originally compiled in Syria for the Ayyubid rulers, was recently translated into English as *Scents and Flavors the Banqueter Favors*. This book contains 635 recipes and medicinal concepts. It presents ways to balance the bodily humors—a medical concept that some scholars believe spread from Europe to East Asia along the early trade routes. The book offers recipes for dishes as diverse as pickled capers, barley vinegar, fragrant flowered waters, and baked goods produced in a clay oven. The range of ingredients is astonishing: almond, apple, apricot, banana, date, citron, cornelian and sour cherries, cucumber, eggplant, grape, hazelnut, lemon, melon, mulberry, orange, pistachio, pomegranate, and quince. The recipes also use asparagus, cabbage, carrot, lettuce, onion, and turnip, as well as herbs and spices such as agarwood, cinnamon, coriander (*Coriandrum sativum*), fennel, garlic, jasmine (*Jasminum* sp.), and other scented flowers, marjoram (*Origanum majorana*), poppy and sesame seeds (*Sesamum indicum*), safflower (*Carthamus tinctorius*), sandalwood, sugar, sumac, and rhubarb.

Another early traveler from China noted a similar abundance along his journey. Departing from central China in the year 1220, by special permission from Genghis Khan, Kiu Chang Chun traveled from Mongolia to the Hindu Kush, into what is now Afghanistan, and back to China. One of his traveling companions, Li Zhichang, wrote down his account of the three-year journey. He observed that when they entered the towns of Central Asia, people came bearing gifts of grape wine and many different kinds of fruits. He discussed cotton and fruit production near the medieval town of Almaty, which he noted was named A-li-ma, from the local word for fruit; he commented on the extensive irrigated apple orchards that surrounded the town.

He discussed rice and vegetable cultivation along the Amu Darya River and fruit orchards in the Tien Shan; he noted peaches, walnuts, and a small peach that might have been an apricot. He particularly praised the fertility of the land around Samarkand, noting that all grains and legumes that grew in China also grew in that region, except buckwheat and soybean. He also praised the watermelons and eggplants—a long, narrow purple variety—of the Zerafshan region.[23] Discussing irrigated farming in the Yin Shan Mountains of Inner Mongolia, he noted that while the fruits ripened late because of the cold in the higher elevations, there were productive irrigated fields and gardens.

A flood of European, Chinese, and Arabic travelers, explorers, merchants, soldiers, and scholars poured into the Central Asian trading cities as the age of colonialism and exploration began; many of them recorded stories of the lively markets and the rich arrays of fruits. In 1671, Ambrosio Bembo (1652–1705), an Italian nobleman, set out on an exploratory mission through the eastern Islamic world, mainly the area of modern-day Iran. He noted the abundance of fruit in every town, especially in Isfahan, and he particularly praised the melons, which surpassed those he knew from the central Mediterranean. He also stopped at a number of caravansaries. He seems to have been largely ignorant of farming practices and likely never worked a plow in his life. Hence he described, with great interest, what a pistachio nut looks like while still on the tree. He threw away the acorns that he was offered as food, because he believed it was beneath his dignity to eat oak nuts, even though the entire caravan he was traveling with stopped to collect them at an oak grove.[24]

Alexander "Bokhara" Burnes (1805–41) traveled to Bukhara through the Hindu Kush on a trip that lasted from 1831 to 1833. A lieutenant in the British army who would become a renowned player in the Great Game, he had been stationed in India as an emissary for the British Crown. Like every other European traveler to pass through Inner Asia, he was astonished by the variety of fruits and nuts available in every bazaar and by the cultivation practices in the high-altitude river valleys of the Hindu Kush. He noted that in every small valley in the mountains, the local people grew apples, cherries, figs, mulberries, peaches, pears, pomegranates, and quinces. He also noted the cultivation practices of the Goklan tribes of Turkmen herders in 1832, recounting that whenever he caught sight of a new camp, "almost every fruit grew in a state of nature. The fig, the vine, pomegranate, raspberry, black currant, and the hazel, shot up everywhere; and, as we approached the camp of the Toorkmuns, there were extensive plantations of the mulberry." Perhaps accustomed to British orchards, with their orderly rows of trees, he seemed unaware that these fruits were not growing wild but were planted and tended by the local herding populations. He referred to Bukhara as the "vegetable kingdom," surrounded by productive fields and orchards.[25] He further noted a lucrative trade in dried fruits from Bukhara across Inner Asia and described apricots, grapes, and peaches drying in the sun. Despite the reputation of the Bukhara grapes, specifically a purple and a long yellow variety, he commented that the oasis was best known for its melons, taking several pages in his log to sing their praises.

James Baillie Fraser (1783–1856) was a Scottish traveler who covered large areas of Central and South Asia on horseback in the early nineteenth century, traveling several routes of the Silk Road with caravans in 1821 and 1822. While most of the epic explorers of the age of discovery tell tales of adventure, excitement, and intrigue, his travel logs are full of melancholy. He concludes that the political situation of the impoverished region of Persia is "despotic, unsteady, and corrupt in the extreme." As he moves further north, he notes that the Turkmen people "are addicted to robbery, murder, and making slaves for sale." In Central Asia he reports massive death tolls from both disease and raiding, including the slaughter of entire villages by Turkmen raiders, a common occurrence before Russian imperial expansion. In fact, only two aspects of the Persian landscape attract his praise: the fruits and the women. Describing a small village in the Semnan province of Iran, populated with displaced Syrians, he notes that the women are very fair, and the only things that vie with the red hue of their cheeks are the apples that grow in the same village.[26] He visits shops in the bazaars, talks to fruit vendors, and stays at caravansaries. He praises several towns for the quality of their fruits, especially in Fergana and along the Zerafshan.

Fraser is also intrigued by the farming practices in the mountains and foothills of Central Asia. In the northern Iran foothills, north of Nishapur, he describes a small village "with plentiful gardens, full of trees that bear fruit of the highest flavour. . . . [These towns] may be seen all along the foot of the hills, and in the little recesses, formed by the ravines whence issues the water that irrigates them." Following the caravan route into the mountains, he notices a glen "finely wooded with walnut, mulberry, poplar, and willow trees; and fruit-tree gardens rising, one above the other, upon the mountain side, watered by little rills that had been led from the stream far above." North of Mashhad in Iran he observes fields of wheat and barley and the cultivation of watermelons, melons, apples, pears, apricots, and a variety of grapes all the way up to the mountains. He notes that almonds, pistachio, saffron, and the "finest" of fruits are exported from Herat, along with locally produced silk. In Central Asia proper, he praises the fruits of the Zerafshan region and states that the "fruits of Bockhara are said to be excellent; apples, pears, quinces, plums, peaches, apricots, cherries, figs, pomegranates, mulberries, grapes, melons, etc. abound in their seasons; the musk-melons are mentioned as remarkable for size and flavour, often weighing twenty pounds, and keeping fresh and good for seven or eight months of the year." He also talks about irrigated farming and crop-rotation cycles in Fergana. He notes farming around Kokand and the cultivation of fruit and nut trees spanning from the city up into the mountains—"lofty pines, poplars, almond, walnut, and pistachio trees."[27]

Eugene Schuyler (1840–90) was an American scholar born in Ithaca, New York; he was also the first American diplomat to visit Russian-occupied

Central Asia in 1873. He visited Khiva, Tashkent, Samarkand, Bukhara, and Kokand. His travel log *Turkistan*, published in 1877, is a vibrant description of the cities and lives of people along the still-active Silk Road. He praises the fruits and nuts and the lavish gardens, orchards, and vineyards of every city that he visits, noting that trays of fresh and dried fruits and nuts are offered to him before meals. Like many other explorers, Schuyler notes that he is continually surrounded by archaeological ruins: the "whole of this region shows traces of ancient cultivation, and it is evident that a very large population at one time existed here. In various parts there are mounds, now covered with growths of saksaul and other shrubs, which are evidently the ruins of former cities."[28] (The appendix includes an excerpt from his accounts.)

Schuyler was accompanied by Januarius MacGahan (1844–78), an American journalist commissioned by the *New York Herald* to report on the journey. He later became famous for his accounts of the Turkish massacres in Bulgaria. Although his travel writing is more concerned with politics and military activities, notably the Russian invasion of the khanate of Khiva, which he observed firsthand in 1873, he too finds time to praise the qualities of the fruits in every city he visited. He notes expansive fruit orchards along the Amu Darya River (which he refers to as the Oxus), with "fruit-trees of all kinds," describing the endless gardens and orchards as a "veritable paradise." He depicts the road up to the city of Khiva as follows: "The apricot trees were still aglow with their golden rosy fruit; miniature rice-fields, still green, were pleasantly varied by yellow stubbles of wheat and barley, now cut and gathered unsheaved in huge stacks like hay, waiting to be threshed by horses' feet." Noting that dried and fresh fruits are transported from Khiva and other Central Asian cities along the great trade routes, he specifically mentions dried fruit from Khiva that is exported to Russia and the cultivation in Khiva of melons, watermelons, cucumber, and arboreal fruits, notably pomegranates and figs.[29] He observes that the Turkmen people, whether living in cities or in farmsteads in the desert, all cultivate plants, and that despite the aridity, the economy is based on farming. (See the appendix for MacGahan's description of the bazaar of Khiva.)

Another traveler in the region was Henry Landsell (1841–1919), an Anglican priest, who conducted multiple expeditions across Asia, including treks into Siberia, the Pamir Mountains, and across Xinjiang. In 1878, he spent time in Semirech'ye in southeastern Kazakhstan and traveled up the Ili River Valley toward China, a region he described as largely under the cultivation of wheat, barley, millets, rye, and oats. In the higher elevations around Semirech'ye, people grew apples, hawthorn, and apricots. Next he followed the Ili Valley into China and the Tarim Basin, noting large melons and perfect apples in the markets of the small towns; he also saw tea, tobacco, sugar, eggs, and a variety of craft goods for sale. He discusses dishes highly spiced with saffron at the restaurants in the bazaar at Kuldja (currently Yinging City; figure 3), twisted and unleavened loaves of bread, a diverse selection of meats (including dog), and, of course, fruits. Shops in the Kuldja bazaar, he noted, were run by vendors from Kashgar, Kokand, Qapal, and Tashkent. Like a proper Anglophile, he was particularly interested in the quality of tea and the methods of preparing it. Traveling higher into the alpine valleys of Inner Asia, he described the farming practices around the town of Vierny (modern Almaty), which included the cultivation of buckwheat and other grains.[30] Schuyler, who also visited Almaty and commented on the orchards flourishing at 2,400 meters elevation, noted that the nearby stream, Almatinsky, was named for the apple trees covering its banks.[31]

Landsell went on to cross the Fergana Plains, noting that they were planted with "fruit-trees of all kinds, 16 varieties of grapes, the usual grains along with rice," as well as cotton and melons.[32] Like so many European explorers before him, he stood in amazement in the bustling Samarkand

Figure 3. Shop street in the town of Kuldja (now Yinging City) in Xinjiang, northwest China. Print from Henry Landsell, *Russian Central Asia: Including Kuldja, Bokhara, Khiva, and Merv* (London: Sampson Low, Marston, Searle, and Rivington, 1885).

bazaar, noting a busy trade in cotton, fruits, rice, silk, and wheat, and he praised the productivity of the farmers of the Zerafshan oases.

Edmund O'Donovan (1844–83), an Irish war correspondent who was killed in Sudan while reporting on the brutal attempts by the British to suppress rebellion in one of their colonies, made a long foray into what is now Turkmenistan through northern Iran in order to witness the Russian siege of the last stronghold of the Turkmen (Turcoman) people at Geok-Tepe. He sat on a hilltop and watched a relatively small Russian force slaughter tens of thousands of Turkmen fighters barricaded in a stronghold that crumbled against a hailstorm of Russian artillery fire. This battle ended Turkmen freedom in 1881. O'Donovan traveled for a considerable distance with a Silk Road caravan. Under Russian dominance, the Silk Road was dramatically changed: the caravansaries were guarded by cannons. He described every town he stopped at as surrounded by orchards and vineyards, watered by irrigation systems that extended for many kilometers. He also commented on the silver trays of dried fruits and nuts that were almost forced on him upon arrival in a town. One small village featured "a pretty extensive grove of trees mostly fruit trees of one kind or another, the jujube [he likely meant *Elaeagnus*], with whity-green foliage like that of the olive, figuring largely amidst the darker tints of the apricot and pomegranate."[33]

He rested at the modern town of Merv (Mary), not far from the ruins of what was once the largest of the Silk Road merchant cities. "Almost throughout the year the bazaar is plentifully supplied with fruits, all of which are of exquisite flavour. In fact, Merv has from time immemorial been celebrated for its fruits. Its melons are occasionally exported to Persia, in which country persons of rank send them to each other as presents."[34] He commented on the different varieties of peaches grown around the city, which were all delicious; his favorite, described as the best he ever tried, was a smaller variety with a deep crimson color. He beamed over the apricots but noted that the Russian olives leave the mouth dry. In the market, he saw Central Asian sun-dried cheese (*qurt*), coagulated milk, mutton, beef, camel meat, and on occasion antelope and wild equid meat, along with plentiful pheasants, other fowl, and eggs. Other market goods included cotton textiles, coarse silk, and woven camel hair; Russian merchants sold long and short rifles, calico, and leather. Other vendors offered green tea and white-lump or crystal sugar; merchants traveling from China brought tea bowls, teapots, and glass tumblers. Others

Figure 4. Photo of a fruit and nut vendor in Samarkand, 1911. The photograph was taken by Sergei Mikhailovich Prokudin-Gorskii, using layered colored plates, on an official mission to document ethnicities of the Russian Empire. Library of Congress Prints and Photographs Division, Washington, DC.

sold food, wooden spoons, dishes, clothing, overcoats, hats, knives, and dried fish. (An excerpt of O'Donovan's description appears in the appendix.)

In addition to strolling the rows of merchant shops, O'Donovan traveled with the Silk Road caravans through northern Iran. He described seeing a mix of cultivated and fallow fields with scattered villages. From the many archaeological ruins he passed, he inferred that there had once been a much greater population in the region. At a protected caravansary in northern Iran, while waiting for the trumpets to sound the departure of a massive caravan heading north to Merv, he described the scene: "Camels and mules laden for the road, with their bells tinkling at every motion, stood around everywhere. The cupola and turrets of Shah Abass's caravanserai stood out boldly against the evening sky."[35] He traveled with that caravan through the towns of Abas-Abad, Mazinan, and Mehrshahr to Sabzavar, where he left it to seek further adventure on foot.

The market bazaar is the center of the social world in Central and southwest Asia. Every city has a large central market square, sometimes open-air and today often covered by a corrugated plastic roof of Soviet manufacture. The bazaar is not just a source of food; it is a node of interaction, social exchange, and commerce. The formation of Arabic and Turkic cuisines gradually took root over centuries of visits to bazaars in numerous cities across the Islamic and Turkic worlds. Figure 4 shows two Uzbek vendors sitting at their booths in the central bazaar of Samarkand in 1911. They are selling apples, lemons (*Citrus* sp.), pomegranates, raisins, dried apricots and prunes, as well as hazelnuts (*Corylus* sp.), eggs, *sushki* (hard, dried bread rings, like pretzels), and melons. The hundreds of varieties of melon grown across Central Asia today are a source of particular pride to the growers, the result of generations of melon farmers cherishing the varieties distinctive to their region or family. Travelers in any Central Asian city in the autumn will be entreated to try the melons, always proclaimed to be the best in Central Asia.

ARCHAEOBOTANIC DATA FOR ARBORICULTURE

Archaeobotanical remains of ancient fruits and nuts illustrate that this productive trade stretched even further back in time than the texts portray. Most of this book will deal with archaeobotanical remains from archaeological sites across Inner Asia. Recently a joint American-Uzbek archaeological expedition into the mountains of Central Asia, directed by Michael Frachetti, discovered a city untouched for a millennium, referred to by archaeologists as Tashbulak.[36] The city was built at roughly 2,200 meters elevation. Excavations at the site in the summer of 2015 allowed the team,

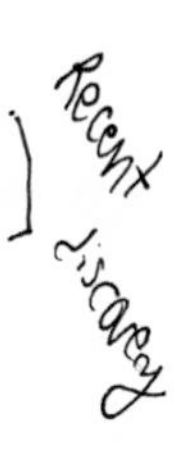

with me as the project archaeobotanist, a view of a Silk Road city market as it would have looked a thousand years ago.

Archaeological surveys in the Zerafshan region have identified dozens of other lost towns scattered through the Pamir Mountains.[37] Some of these ancient towns are located along a vein of exposed metal ore that runs through Uzbekistan and Tajikistan. The mining towns range in date: the earliest, the ancient city of Sarazm, was founded in the late fourth millennium BC, but others were established as recently as the Soviet period. Located at elevations between 2,000 and 5,000 meters, these towns would have required a steady supply of food from other locales to feed the population. Some of these towns may represent the earliest entrepôts of established trade routes. Their metal ores and smelted metal goods supplied the growing empires in South Asia, thus representing the earliest organized trading system on the great Silk Road. The mining supply routes may constitute the origins of the Silk Road.

As a member of the excavation team at Tashbulak, I assisted in excavating a small (2 × 1 meter) trench directly into a large midden (garbage) deposit in the center of the ancient town, at the site of the central market bazaar. The sediment samples from this midden have yielded remains of carbonized and preserved apple seeds (see figure 23), peach and apricot pits (see figures 25 and 26), grape seeds (and even a whole grape; see figures 20 and 22), melon seeds, pistachio shell fragments, rosehip seeds (*Rosa* sp.), Russian olive pits, hackberry pits (*Celtis* sp.), and cherry pits, as well as peas, chickpeas, and wheat and barley grains.[38]

Most of this produce was likely not grown in or near the town. Most fruit crops, especially arboreal fruits, would have produced limited harvests at that elevation, with a short growing season, but a few hours' journey away, at lower elevations, were areas that could have supported orchards and gardens. Many of the products sold in Tashbulak might have traveled significant distances. The market was likely a hub of communication, connecting the inhabitants of the town with an Islamic population extending well beyond what is now Samarkand.

A Soviet-run excavation in 1986 at another high-elevation mining center, roughly contemporary with Tashbulak (about a millennium old), has provided additional knowledge about the goods for sale in these towns. The archaeological site of Bazar-Dara (Cliff Market) is located at an elevation of nearly 4,000 meters on the bank of the Ak-Dzhilga River in the larger Murghab River valley of southeastern Tajikistan, near the Badakhshan region of Afghanistan. This mining town consisted of roughly eighty architectural structures.[39] When the Soviet archaeologists excavated the ancient houses, they recovered an astonishing variety of perishable goods, including grains and legumes as well as many fruit pits and nut shells; these items had been preserved because of the cold climate at this high elevation, like food in a freezer. The plant remains, while not systematically collected by the team, led to the largest archaeobotanical study in Central Asia. The fruit remains included apple and pear seeds, apricot pits, barberry seeds (*Berberis* sp.), cherry pits (*Prunus* sp.), grape pips, melon seeds, mulberry seeds, peach pits, and watermelon seeds (*Citrullus lanatus*); the nut remains included almond, hazelnut, pistachio, and walnut shells. The excavators also claim to have recovered remains of a few fruits of greater interest, such as date pits and persimmon (*Diospyros* sp.), and, astonishingly, one possible coconut-shell fragment.[40] I have not, however, been able to track down the remains of these ancient fruits, and no full descriptions or photographs are included in the report.

The rich diversity of fruits and breads that Babur, Olearius, and Abul Fazl tasted while passing through the oases that dot the Central Asian deserts with green is a legacy of the ancient Silk Road. Simple merchants, migrants, exiles, or vagabonds who passed through these centers carried in their satchels or saddlebags seeds, fruits, roots, clippings for propagation, and tree saplings from one end of Asia to the other.

While these merchants are mostly faceless and nameless to us today, a few legendary characters stand out. For example, historical sources from the court of the Assyrian king Tiglath-Pileser I (believed to have reigned from 1115 to 1102 BC) state that the king took credit for discovering several trees, including a cedar and an oak.[41] Many sinologists credit the mythicized figure of Zhang Qian with introducing a wide array of crops to China, although Berthold Laufer, a prominent historian of Silk Road trade, stated that "Čań K'ien . . . brought to China solely two plants,—alfalfa and the grape-vine. No other plant is attributed to him in the contemporaneous annals."[42] Earlier archaeological finds of grapes in China suggest that Zhang Qian did not introduce even these two crops.

At the other end of the Silk Road, Alexander the Great is often credited with introducing to European cuisine many crops that he encountered in his eastward expansion of the Macedonian Empire, most notably apples—particularly dwarf variants. There is little solid evidence, however, to credit him with any specific crop discoveries. Dioscorides is similarly credited with the discovery of crops during his travels with the Roman army. While it is unlikely that he discovered any new plants, he probably did spread the knowledge of many crops through his writings.

Perhaps the most famous figure associated with the spread of foods along the Silk Road is Marco Polo. His account of his journey from Venice to China has not only been exaggerated over time but also seems to have been significantly embellished by Polo himself. The tales are so lively, and so hard to reconcile with historical sources, that some historians have questioned whether his journey ever took place at all. In the late 1990s, Frances Wood published *Did Marco Polo Go to China?*[43] More recently, Hans Ulrich Vogel responded with a book titled *Marco Polo Was in China*.[44]

Whether factual or mythical, the story recounts the journey of the young Marco Polo from Venice to China, where he purportedly served at the court of the Mongol emperor Kublai Khan. Polo, born in 1254, claimed that he left

Italy at age seventeen with his father Niccoló Polo, his uncle Maffeo Polo, and two Dominican monks, and did not return for twenty-four years. Niccoló and Maffeo, as Silk Road merchants, had purportedly made the same journey once already, between 1260 and 1269.

Polo is said to have returned to Venice in 1295, but he was caught up in a naval war between Venice and Genoa, both powerful city-states and mercantile ports, over rights and access to markets for spices and crops across the Mediterranean. While imprisoned in Genoa, Polo recounted his journey to a fellow inmate, Rusticello from Pisa, whom history credits with writing down the tale. Polo lived out the remainder of his life as a wealthy merchant in Venice, and his legend has been passed down for generations. But even if Marco Polo's tales were fabricated, they must have been based on stories from other merchants who had traveled along the Silk Road.

The myth that Polo was responsible for introducing pasta to Italy is certainly false. His account does refer to millets, rice, and other crops in China, as well as mulberry trees, grown to feed silkworms and produce fiber for paper.[45] And since he does describe noodles, it is possible that he was already familiar with them. When he describes the wheat noodles produced in the dynastic center of China, he calls them vermicelli, lasagne, and *lagana*, suggesting that he was comparing them to dishes with which he was already familiar.[46] Because the original text does not survive, however, this assertion cannot be proved. The myth that Polo introduced noodles to Italy was most likely first popularized by an article in the 1929 issue of *Macaroni Journal*, the official trade journal of pasta manufacturers.

The historians who argue that noodles originated in the Mediterranean point to the unleavened flatbreads of the ancient Greeks, called *laganon*. They suggest that eventually people started boiling this flatbread and adding layers of cheese, creating lasagna, and that the boiled flatbread was also cut into strips called *itria*.[47] This culinary technique could have spread along the early Silk Road, reaching China before the foundation of the Tang Dynasty in the seventh century AD. This theory is plausible, but most scholars subscribe to the idea that noodles were introduced to Italy by Arab merchants—leading to the invention of ballerine, bucatini, cannelloni, capelli d'angelo, cocciolette, farfalle, fettuccine, fusilli, lasagna, linguine, macaroni, penne, perciatelli, ravioli, rigatoni, spaghetti, tagliatelle, tagliolini, tortellini, and ziti, to name just a few of the pasta shapes popular in Italy and around the world today.

The only archaeological discoveries of proper macaroni wheats in Central and East Asia (hard wheat, durum, or semolina; *Triticum turgidum* ssp. *durum*) that I made were at the medieval site of Tashbulak. Some historians have claimed that they spread with other novel crops during the Islamic conquests in the seventh century AD.[48] Chinese noodles are made from a variety of grains, but modern Mediterranean pastas tend to be made primarily of hard wheat (although this may be a more recent trend, resulting from the mechanized milling and sieving of fine flour). As much as Italians may wish to claim pasta as part of their national heritage, that claim is hard to reconcile with the lack of any unambiguous mention of noodles or hard wheat in the ancient Roman or Greek texts—especially when the surviving literature includes detailed descriptions of entire dinner parties and whole recipe books. The earliest clear references to noodles in literary sources from southern Europe date to the twelfth and thirteenth centuries and appear to link the spread of noodle making to Arabic traders.[49] It is also safe to say, based on literary sources, that noodles were being consumed in southwest Asia many centuries earlier. Therefore, it is likely that this quintessentially Italian dish came to Italy from Asia less than a millennium ago, introduced by seafaring Arab merchants.

Ultimately, the only way that we can separate reality from myth is by comparing evidence from historical sources with data from archaeological investigations. Of course, the archaeological record is not free from error or misinterpretation: as modern scientific methods are more widely adopted in archaeology, earlier archaeological interpretations are being called into question. The data that I present in this book result largely from archaeobotanical studies. However, isotope analysis, ancient proteomics, and ancient genetics studies are now increasingly complementing the results from archaeobotanical research. The chapters that follow contain the conclusions of numerous multidisciplinary scholarly studies, collectively providing a robust image of the past.

CHAPTER THREE

The Silk and Spice Routes

When Vasco da Gama's fleet of four ships and roughly 170 crewmen landed in Calicut, India (modern-day Kozhikode), on May 20, 1498, it marked the first successful all-sea voyage from Europe to East Asia. Da Gama's passage opened a new chapter for the Spice Route, connecting Europe to southeast Asia and leading to the distribution of new flavors, notably black pepper (*Piper nigrum*) and cinnamon (*Cinnamomum verum*), around the globe.[1]

His voyage was one of several attempts to reach Asia by sea. Six years earlier, on March 4, 1493, Christopher Columbus's fleet had pulled into the same Lisbon harbor after being diverted by a storm on its return from the first of his transatlantic explorations. This voyage altered the course of human history and initiated the vast Columbian exchange of living things between Europe and the New World. Over the following centuries, Europeans would learn about avocado (*Persea americana*), squash (*Cucurbita*), cacao (*Theobroma cacao*), cassava (*Manihot esculenta*), common beans (*Phaseolus vulgaris*), chili peppers (*Capsicum annuum*), maize (*Zea mays* ssp. *mays*), peanuts (*Arachis hypogaea*), pecans (*Carya illinoinensis*), potatoes (*Solanum tuberosum*), quinoa (*Chenopodium quinoa*), sunflowers (*Helianthus annuus* ssp. *annuus*), tobacco (*Nicotiana rustica* and *N. tobaccum*), tomatoes (*Solanum lycopersicum*), vanilla (*Vanilla* sp.), and wild rice (*Zizania aquatic*). In exchange, they introduced to the Americas a wave of plant species so invasive that it has reshaped every ecosystem in the Western Hemisphere.

To Columbus's great disappointment, he never found the Asian spice plants that he had set out in search of, such as mace (*Myristica fragrans*), nutmeg (*M. fragrans*), and black pepper.[2] In his personal log on Tuesday, October 23, 1492, he wrote, "That I have no personal knowledge of these products causes me the greatest sorrow in the world, for I see a thousand kinds of trees, each one with its own special fruit, all green now as in Spain during the months of May and June, as well as a thousand kinds of herbs with their flowers; yet I know none of them except this aloe."[3] In their ignorance of the native plant life, several of Columbus's men almost died after eating manzanillo fruit (*Hippomane mancinella*, known in Spanish as *manzanilla de la muerte*, "little apple of death"), thinking that the round green fruits were the familiar apples of Portugal.[4]

Both Columbus and da Gama had sound reasons for setting sail from their safe European harbors into *mare incognitum*. They were both searching for an efficient maritime route to southeast Asia, the source of the goods of the legendary Spice Routes. As with the Silk Road, exchanges along the more southerly Spice Routes can be traced back at least to the third millennium BC. By the first millennium BC, much of this exchange was switching over to a largely nautical route. Goods were brought up the Red Sea and then carried overland to the Mediterranean or across to the Nile River and north to Mediterranean ports. Among the spices being carried by Persian merchants to Rome by the first century AD were black pepper, cardamom (*Elettaria* and *Amomum* spp.), ginger (*Zingiber officinale*), and turmeric. The Axumite Empire (AD 100–940) dominated the Red Sea trade until the Islamization of the region, when Arab merchants took control of it.

The mission of Ferdinand Magellan (1480–1521) sought a sea route to southeast Asia to complete the mission Columbus had set out to accomplish. During its three-year voyage, the Magellan expedition lost four out of five ships and two hundred crew members, including Magellan and most of his officers, but it succeeded in its aim. The surviving ship, the *Victoria*, captained by Juan Sebastian Elcano, not only was the first to circumnavigate the globe but also brought back a cargo of treasure: its crew returned to Seville in 1522 with a hold full of cinnamon, mace, nutmeg, and twenty-six tons of cloves (*Syzygium aromaticum*) from Ternate in the Maluku Islands, the core of the Spice Islands of Indonesia. Their cargo more than repaid the Spanish crown for its investment in the voyage.

Plant dispersal along established vectors of human contact, such as the Atlantic shipping routes or the spice routes, was by no means an innovation of the colonial period. Much earlier routes of botanical exchange include the Axumite salt trade route, along which caravans of dromedaries carried salt across large expanses of the Ethiopian deserts, and the Persian Royal Road, which crosses the arid regions of southwest Asia. Additionally, the Sabean Lane, also known as the Sabian or Sabaean Way, was the route by which crops of Indian and East Asian origin spread into northeast Africa and southwest Asia, and vice versa, along the southern periphery of the Arabian Peninsula.[5] Crops such as finger millet (*Eleusine coracana*), sorghum (*Sorghum bicolor*), pearl millet (*Pennisetum glaucum*), cowpea (*Vigna unguiculata*), castor bean (*Ricinus communis*), citrus, and possibly sesame traveled this route during the third and second millennia BC. Archaeobotanical discoveries over the past couple of decades have shown that many of these crops originated in North Africa and spread eastward.[6] For example, archaeobotanical remains of pearl millet in the Sahel region of North Africa have been dated to 2500 BC, and the crop appears to have spread to India by 2000 BC.[7] Sorghum followed the same route by the early second millennium BC.[8] The archaeobotanical evidence of these crops in India is complemented by references in Sanskrit and Vedic texts. While these Arabian spice routes were likely in place as early as 2800 BC, from 1200 to 650 BC they were controlled by Minaeans from the land of Punt—a little-known nation that has provoked considerable debate among historians—who were major trading partners of the Egyptian Empire.[9]

During the first millennium BC, the Sabeans dominated trade in Arabia and the southern parts of southwest Asia. They traded in frankincense and myrrh from North Africa, which were used by the Greeks and Romans to honor their gods and venerate the dead. The Sabean traders traveled into Africa, Asia, and Eastern Europe. Sabean farmers used a complex system of canals to water hundreds of square kilometers of fields and orchards.[10] The system largely relied on a 1.6-km-long main canal, transferring dammed water from the Wadi Adhanah through a system of secondary and tertiary canals. The harvest of grains, legumes, and fruits was carried by horses or camels to be traded with the Minaeans for frankincense, myrrh, and fennel.[11] Thus the Sabean farmers of the Ma'rid Oasis and Minaean traders forged the mercantile enterprise that would eventually evolve into the great Spice Routes, setting tens of thousands of European ships to sea and launching the age of exploration and colonialism.

The Sabean Lane is also known as the Incense Road or Frankincense Trail. These two terms refer to different kinds of plants. The Incense Road connected the eastern Mediterranean with India through Somalia, Egypt, and the Levant, northeast Africa, and Arabia by means of the Horn of Africa, and trade along this route peaked between 700 BC and AD 200. It was the early spice trade of the Arabian Peninsula, specializing in cardamom, Damascus rose (*Rosa damascene*), pomegranates, sumac (*Rhus coriaria*, used in the spice mixture *za'atar*), and turmeric.[12] In addition to Asian spices, textiles, and precious stones, the trade included Arabian frankincense (*Boswellia* spp.), bdellium (*Commiphora wightii* and *C. africana*), ebony (*Diospyros* spp.), and myrrh (*Commiphora* spp.). Frankincense, bdellium, and myrrh are all aromatic tree resins, and they originate mainly from North Africa. By the time the Sasanian Empire came to power in Persia in AD 224, new crops were moving across Arabia, such as Asian rice, bananas and plantains (*Musa* sp.), eggplants, spinach, citrus fruits, and sugarcane (*Saccharum officinarum*).[13]

The Islamic conquest of Persia in the seventh century saw even farther-reaching exchange, bringing novel crop varieties and culinary traditions to southwest Asia. As the nautical spice routes became increasingly prominent, more spices with distant origins, such as black and white pepper, cassia (*Cinnamomum cassia*), cinnamon, nutmeg and mace, star anise (*Illicium verum*), and cloves reached the Mediterranean.

The center of the southeast Asian spice trade was the port of Malacca, where the streets were covered in the dust of pepper and nutmeg. In the sixteenth century, the Banda Islands were the only place on earth producing nutmeg and mace. Cloves came from Malacca and southern Malaysia. At the same time, the port of Malacca on the Malay Peninsula was flooded with ships carrying Portuguese guns and supplies from Europe. Europeans jousted for power in the Indian Ocean for the entire sixteenth century: the nation that controlled the spice trade controlled the greatest flow of wealth in the world.

However, the spice trade did not start with the introduction of European colonial powers. People had been moving these plant products over great distances for millennia. The peoples of India played a significant role in exchange along both the Silk Road and the Spice Routes. In the first millennium BC, north India and Pakistan (Uttarapatha) was the hub for exchange, connecting trade routes through the Indo-Gangetic Plain in the northern part of the subcontinent with ports on the Bay of Bengal.[14] Other routes from Uttarapatha stretched out toward the Ganges River. By the end of the Mauryan Empire (322–185 BC), the river had become a highway for trade and commerce. Historians have suggested that some of the Indian trade routes were formed by the early migrations of Vedic Hindus, Buddhists, and Jains out of the Ganges region and across South Asia.[15] Historians also claim that the Kushan Empire (AD 78–144), extending over northern India and southern Central Asia, controlled exchange along the routes to the north that became the southern Silk Road. These interweaving trade paths fed into larger arteries of the Silk Road, meeting up with caravans that were traveling from Xi'an to Kabul or Mashad, or from Bukhara, Khiva, Merv, Paykent, or Samarkand to Rome or to the Indus. Individual caravans rarely traveled through more than a few of these trading towns, as goods were passed from one merchant to another. At the trading town of Palmyra in Parthia, a number of preserved textiles were recovered by archaeologists: one silk fragment, woven in Chinese style, depicts a scene of grape harvesting, with Central Asian merchants and Bactrian camels in the background.[16]

Many historians have focused on the goods moving westward, connecting these regions to southern Europe. During the Roman period, spices and precious stones were being transported by Kushan traders from Kashmir and the broader Himalayan region; among these spices were costus (*Saussurea costus*, Greek for "from the east"), bdellium, and Himalayan spikenard (*Nardostachys jatamansi*). Turquoise was carried from Khorasan in northeastern Iran and lapis lazuli from Badakhshan in northeastern Afghanistan. Many of the merchants on the Red Sea were descended from Greeks who established themselves in the region during the Ptolemaic Dynasty. After the fall of the Kushan Empire, the southern Silk Road was left in the hands of small trading networks and merchant groups. The Sogdians, from modern-day Samarkand, emerged as the dominant trading community. In the third century, people in the Sasanian Empire supplanted the Parthians as rulers of the Persian Empire and pushed out the Kushan, who were further displaced in the fourth century by the Gupta Empire (AD 320–550).[17]

In this book, I focus on the overland trade routes, but once merchant ships started moving goods across Asia, the rates of exchange increased greatly. The terminus of maritime routes of the southernmost branch of the Spice Routes was in Zayton (modern Quanzhou). Its traffic peaked in the last quarter of the fourteenth century, after Timur had completed his conquest of Central Asia, including modern Armenia, Azerbaijan, Iran, Iraq, and Georgia. Timur established his capital in Samarkand and changed the political atmosphere of the exchange routes. A year before his death on February 18, 1405, he attempted to expand eastward as well, rallying a large-scale military attack against the Ming Dynasty in China. While this campaign failed in its primary object, it did succeed in diverting Ming mercantile interests away from Central Asia. As the Ming turned their attentions toward nautical routes to the south, the ancient city of Zayton became a burgeoning trading hub, dominated by Islamic merchants, connecting East Asia to Europe. Marco Polo supposedly visited Zayton in 1292. By the late fourteenth century, a steady stream of merchant ships were loading black pepper, cassia, cloves, nutmeg, mace, Sichuan peppercorns (*Zanthoxylum* spp.), and sandalwood (*Santalum album*) in Zayton for transport to distant markets in the west.

Shipping ledgers from Zayton dating back to the thirteenth century have been analyzed by the ethnobotanist Gary Paul Nabhan. They list aloe, apricots, betel (*Piper betle*), black cardamom (*Amomum subulatum*), cassia, cinnamon, cloves, coconuts, coriander, cumin, dragon's blood, fennel, fenugreek (*Trigonella foenum-graecum*), frankincense, ginger, green cardamom (*Elettaria cardamomum*), hazelnuts, hemp seeds, myrrh, osmanthus (*Osmanthus* sp.), pepper, pine nuts, rhubarb, saffron, sandalwood, sapanwood (*Biancaea sappan*), and star anise.[18]

While the Spice Routes played a profound role in shaping cuisines, arguably the more northerly Silk Road did more to deliver many of the ingredients in our kitchen cabinets. The exponential intensification of these networks of communication and the resulting global dissemination of ideas and technology transformed human history and continue to push us into the Information Age.

BRIDGING THE ANCIENT WORLD

The Silk Road was not a single road or even a set of regular paths: it is better thought of as a dynamic cultural phenomenon, marked by increased

Figure 5. The fruit and vegetable market in Bukhara, just next to the old city wall, 2017. Today's vendors sell many of the same arboreal crops that would have been sold in this city two millennia ago. Photo by the author.

mobility and interconnectivity in Eurasia, which linked far-flung cultures. As such, it extended north through the Altai Mountains into Siberia and Mongolia; it brought goods northward from Africa, the Arabian Peninsula, and India; and, of course, it connected the Mediterranean and East Asia to Central Asia.[19] This network of exchange, which placed Central Asia at the center of the ancient world, looked more like the spokes of a wheel than a straight road. Central Asian porters carried innovations from China to Tibet, Bactria, Persia, Byzantium, Greece, Rome, and beyond.

This reticulated network may have originated from routes used by herders and agropastoralists of the second millennium BC. In weaving together decades of Soviet scholarship on transhumance in the foothill regions across Inner Asia, E. Kuzmina notes that the herders' routes through the mountains crossed passes where water and forage were available and the slopes were not too steep. Hence, she argues, by "tracing the origins of their operation, it can be inferred that the precursors for the Great Silk Road routes began as far back as the Bronze Age."[20] This idea is prominent in the Soviet archaeological literature.[21]

The term *Silk Road* (or, more accurately, *Seidenstrasse*) was coined in 1877 by the explorer, geographer, and historian Baron Ferdinand von Richthofen, the uncle of the legendary Red Baron pilot of World War I. He was drawing on Classical accounts of the origins of silk in some unknown land to the east. David Christian and some other historians prefer the plural term *Silk Roads*, claiming that the term comes from the German plural *Seidenstrassen*.[22] I use the singular here in keeping with convention and familiarity.

Although Richthofen applied the term *Silk Road* in a very narrow sense, he was also interested in broader exchange systems connecting Eurasia after the Han Dynasty, especially the networks bringing exotic goods into China during the Tang.[23] He realized that by the time of the Tang Dynasty, there were vibrant exchange networks. As James Milward summarizes, "The narrow concept of the silk road as an east-west route between China and Rome likewise obscures the fact that there was not one 'road' but rather a skein of routes linking many entrepôts. Historians think of the silk road more as a network than a linear route."[24] Similarly, the historian Colin Renfrew notes that there were "many Silk Roads, yet their main highways extended via Xinjiang through Inner Asia and on to Syria on the Mediterranean coast and so to the Classical world."[25] Over several millennia, Central Asians and their trade networks shaped the modern world. The many labels that historians have applied to their societies and cultures include Avars, Cimmerians, the Golden Horde, Khotanese, Mongols, Saka, Sarmatians, Scythians (broadly speaking), Sogdians, Tocharians, Uighurs, Wusun, Xiongnu, and Yuezhi.

Archaeological research continues to show that the Silk Road did not just appear: rather, it gradually formed out of seasonal mobile pastoralist herd movements and high-elevation mining town supply routes. The Russian historian and Silk Road scholar Natalya Gorbunova builds on the idea that the Silk Road routes developed gradually as a result of seasonal migration, and especially the increased reliance on horse transport in the mountains of Central Asia by the late first millennium BC.[26] Thus the camel caravans of the late first millennium AD may have been following tracks worn by sheep and goat herders centuries earlier. Likewise, the mountain passes guarded by Tibetan yak herders and Turkic raiders were used by pastoralists for millennia to travel between winter and summer pastures.

Following a growing trend among scholars, I use a broad definition of the Silk Road, one that goes beyond the familiar trope of long-distance caravan exchange during the Classical and Han dynastic periods.[27] This traditional, romanticized image envisions politically organized, imperial groups connected by long-distance exchange. In this view, the great road originated only in the first millennium AD and lasted until the introduction of the combustion engine. Yet the caravans continued into the modern era, and the famous ethnographer and explorer Owen Lattimore famously rode with them while traveling in Central Asia, as described in his book *The Desert Road*

Figure 6. Herd of dromedaries between oases, Kara-Kum Desert, Turkmenistan, 2010. There are few large camel herds left anywhere in Central Asia; they have mostly been replaced by trucks and airplanes. Photo by the author.

to Turkestan. During the 1920s and 1930s, he traveled through markets and bazaars in China, Mongolia, Russia, and Central Asia.[28]

As people gradually dispersed through the narrow mountain passes, they carried with them the knowledge of and equipment for advanced forms of metallurgy, horse breeding and riding, textile and craft manufacturing, and religions. They carried textiles of silk, wool, linen (*Linum usitatissimum*), hemp, and cotton (*Gossypium herbaceum*), woven with tall, warp-weighted looms on which they could produce complex twill patterns and plaids.[29] They carried furs and felts, cut and polished stone ornamentation, ceramic vessels and art, salted fish and meat, and sun-dried cheese, dried fruits, and fermented beverages such as grape wine and *kumiss* (fermented horse milk). They spread languages, genes, and a wide repertoire of cultural practices.

As technology and cultural knowledge were disseminated along the route, it also became a conduit for other kinds of knowledge. The collapse of the Greco-Bactrian Empire in 130 BC was the first historical event to be narrated by historians at both ends of the Silk Road.[30] Between 109 and 91 BC the legendary Chinese historian Sima Qian and his colleagues wrote down the tales of Zhang Qian, the Han Dynasty military envoy who was sent to try and make an alliance with peoples in Central Asia against the Xiongnu. While in Central Asia, Zhang learned about the fall of the Greco-Bactrian nation several years earlier, and thus it was recorded in Sima Qian's account.[31] Roughly a century and a half later, the classical historian Strabo scratched onto his scrolls or wooden slabs the known history of Central Asia and, likely referencing earlier sources, noted the collapse of the Greco-Bactrian Empire and the northern expansion of the Yuezhi.[32]

We have firm evidence of the flow of trade along the Silk Road in the first century AD, with the regulation of trade and organized taxation. Shortly after 121 BC, Emperor Wudi (Wu of Han), who commissioned the expedition of Zhang Qian, extended the Great Wall all the way to Dunhuang and the Gate of Jade frontier post (Yumenguan), which became the westernmost of the Han military forts. The ruins of this fort still sit about eighty kilometers northwest of Dunhuang.[33] As a result of the establishment of military posts, caravan cities formed in western China, notably along the Hexi Corridor, a pass through the Qilian Mountains leading to Chang'an on the west bank of the Yellow River.[34] Private and military documents written on wooden slabs have been recovered from Han military guard towers in the Western Regions. Around 1900, the archaeologist Aurel Stein recovered similar documents from garbage piles at watchtowers near the Thousand Buddhist caves, along the route from the ancient gate of Cathay. Among these documents, which were tied with silk strips, is one that lists chores assigned by the military leader of the watchtower: cultivating fields, raising garden vegetables, digging canals, and repairing domestic tools.[35]

The Han military incursions into the region brought a wave of agricultural innovation, especially in the Taklimakan Desert, which increased the population and supported more exchange.[36] The Han Dynasty conquered the area of the Kororaina in 108 BC and set up military outposts in the region. Historians have suggested that the mingling of the Chinese military with local peoples led to a greater popularity of grapes, pears, pomegranates, and dates in the Han Empire.[37] But it was really the expansion into northern Central Asia by the Han that led to the flow of Central Asian foods eastward.

In the early first millennium of the common era, the Silk Road began to take on a very different appearance. The Tuoba (Tabghach) people conquered the Han in AD 386 and loosely held it as the Wei Dynasty until around 550. The Tuoba, who were from the north and are often characterized as a nomadic group, maintained close cultural and trade ties with Central Asians. Some historians argue that the period from 400 to around 500 was among the most important for exchange along the northern routes, nurturing the Tang elites' great hunger for exotic goods. The capital of the Wei Dynasty, before its collapse, had a foreign quarter (as did the Tang capital later), whose population some historians estimate at forty to fifty thousand Central Asians.[38] By the 600s the Tang Empire was strongly multicultural, and markets were teeming with Central Asian merchants selling wine, fruits, dried meats, and horses.

One of the best ancient texts for understanding the introduction of new foods from the west during the Wei Dynasty is the *Qimin Yaoshu* (Essential techniques for the welfare of the people), thought to have been completed in AD 544, as Wei control was waning. The work is attributed to Jia Sixie, and there are several sections dedicated to agricultural practices; along with crops such as Central Asian melons and cucumbers, the book mentions pomegranates, which appear to have been newly introduced.[39]

By the end of the Wei Dynasty, certain Sogdian families, originally from the foothills of the Pamir Mountains and the rich floodplains of the Zerafshan River, were starting to take on the prestigious role of the middlemen of the Silk Road. These merchants crossed Central Asia on perilous treks to bring silk and spices to distant markets. The Bactrian (two-humped) camel was stronger, able to carry larger loads, and more tolerant of the harsh environments of

Central Asia than its Arabian cousin, the dromedary, and the classic images of the Silk Road, with their long camel caravans, have their origins in this era. Despite the hardiness of the animals, however, the most vital element of the Silk Road was its trading posts, which served as refueling stops. These oasis and mountain valley towns, such as Bukhara, Loulan, Merv, Samarkand, Turfan, and Urumqi, have all left their mark on human history.

The Tang cultivated the greatest trade network the world had seen, creating an increasingly cosmopolitan atmosphere. Life in the palaces and the market squares was enlivened by drinking parties with wines imported from the Western Regions (the oasis cities and river-valley villages in parts of Central Asia and the area that is now Xinjiang). These festivities were accompanied by music and whirling dancers from Central Asia. Exotic male and female performers were imported, along with waitresses to serve the wine.[40] The dynastic capital of Chang'an housed numerous wine shops, notably near the western market and in the Liquan ward of the city.[41] These shops, often owned by Sogdians, were centers for cultural exchange and mercantile negotiations. Historical sources identify Central Asian influences, noting waitresses with blue eyes and Persian vendors selling breads in the streets.[42] During this time, Chang'an became the world's largest metropolitan center as well as probably its most diverse. The city was home to adherents of Buddhism, Islam, Judaism, Nestorian and Orthodox Christianity, and Zoroastrianism; there were Arabs, Han Chinese, Indians, Mongols, Persians, Sogdians, Tajiks, Tatars, Turkmen, and Uighurs. These people were not just mingling in the markets: they were forging trade and marriage alliances. Thus the Silk Road was a conduit not only for tea, musk, and horses but also for human genes.

The eminent sinologist Edward Hertzl Schafer compiled an exhaustive study of Tang Dynasty imports based on historical sources. They included humans—both slaves and performers—and domesticated animals, notably horses. They also included a wide range of wild animals and animal products, including ivory and rhinoceros horn; furs from animals ranging from martens to leopards; and tails, horns, and feathers. Secular and sacred objects were transported, as well as books and texts, precious and utilitarian metals, and processed glass. The stones and minerals imported included jadeite and nephrite, carnelian, realgar, quartz, malachite, lapis lazuli, pearls, turquoise, amber, coral, salt, alum, borax, saltpeter, sodium sulfate, sulfur, and diamonds. The merchants traded wool, linen, silk, and cotton textiles as well as felts, rugs, and garments.

Schafer lists a few of the numerous drugs, herbs, remedies, and mythological treatments that were traded; he also mentions aromatics, including incense, agarwood (*Aquilaria* and *Gyrinops* spp.), lakawood (*Dalbergia parviflora*), elemi resin (*Canarium luzonicum*), camphor resin (*Cinnamomum camphora*), styrax or sweetgum (*Liquidambar orientalis*), frankincense and myrrh, cloves, patchouli (*Pogostemon cablin*), and jasmine. Many preserved foods were transported, including wines and other fermented beverages, dried fruits and vegetables, and cane sugar (*Saccharum officinarum*).[43] Other historians have listed a range of products that moved along the great road during the Tang period, as attested by Arab scholars.[44]

Gold and silver artifacts recovered by archaeologists in China attest to the prominence of trade and the spread of ideas during the Tang Dynasty. One notable find is the Hejia Village hoard. In the 1970s, two large ceramic vessels and a smaller silver pot were dug up a kilometer southeast of the former western or foreigners' market of Chang'an. The two pots, each roughly half a meter tall, were filled with gold and silver goods, medicine bottles, and carved precious stones and minerals. In addition, a large collection of coins was recovered, some originating from as far away as Japan and Byzantium. They included a solidus, or gold coin, apparently from the reign of Heraclius (AD 610–41), although it may be a Chinese-made ancient replica, and a Sasanian drachm from the reign of Khusrau II (AD 590–628).[45] Collectively, the 478 coins and the dates on several silver ingots dated the hoard to sometime shortly after AD 732. In addition to the gold and silver items, the find included agate vessels, carved jades, topaz, sapphires, amber, and mountain coral (a red marine coral, *Corallium*, highly prized as a gem in Tibet). Some of the vessels presumably held medicines and other precious trade goods. Many of the vessels, notably wine-drinking bowls, are made in typical Sogdian styles (although, again, they may be replicas produced at the foreigners' markets in China), decorated with hunting scenes, drunken parties with dancers and Sogdian musicians, birds carrying ribbons, lions, and the characteristic Sogdian pearl borders.

The Tang connection with Central Asians began to erode with the Arab expansion eastward. After fierce battles and a long siege in 712, Samarkand finally fell to the Arab invaders, who had already overrun southern Central Asia and Iran. The Arab forces were led by Qutayba b. Muslim. Ghurak, the king of Sogdiana, signed a treaty of surrender and paid a large ransom for the city. Arab leaders occupied the city, however, forcing the Sogdian leaders to move to nearby Ishtikhan, and control of the Silk Road trade passed to a new regime. In light of the strong connections between the Tang and the Sogdians (as well as other peoples of Central Asia), the Tang Dynasty sent a large military force into Central Asia to support local political leaders, who still hoped to repel the Arab invasions. In July 751 Tang military forces went up against the Abbasid forces of the expanding Arab army, under the leadership of General Ziyad b. Salih, in the infamous battle of the Talas River.[46] The Islamic army defeated the Tang army and took a large number of prisoners, including many artisans and skilled workers. According to legend, these artisans had the ability to weave reeled silks. Mulberry trees were planted across the realm of the caliphate to support silkworms, and silk was soon being produced across Asia. (Some scholars have suggested, however, that there may have been earlier silk industries in Central Asia.)[47]

Over the next millennium, control of trade along the Silk Road passed from one military power to another. The legacy of the battle of Nahavand

(642), when Arab forces pushed the Sasanian armies back to Merv, paved the way for the Islamization of the Silk Road. Archaeologically, there is little evidence for a violent Islamization of Central Asia, and in many areas the cultural shift may have been gradual. The Abbasid Caliphate (750–1258), which replaced the Umayyad Caliphate (661–750), moved the capital from Damascus to Baghdad. The Abbasid elites accumulated considerable wealth, notably in rugs and silks from China and India. Many of the stories collected in the *Arabian Nights* date from this period. Organized trading networks were established and protected by the caliphate by the tenth century, and these routes extended into North Africa, down the east coast of Africa, and even into some areas of West Africa, not to mention most of Europe and all of Asia. The major posts along the trade routes were largely Islamic, even in regions where Islam was not the major religion, because trade was dominated by Muslim traders. These Islamic institutions quickly supplanted their Buddhist predecessors as the holders of power in Central Asia. The unified Seljuk Sultanate and independent emirs secured the western parts of the Silk Road, ensuring safety for the caravans by building fortified caravansaries that provided food and lodging for travelers and fodder for their animals.[48]

During the Golden Age of Islam (750–1257), a grand fusion of culinary traditions of Persian and Arabic origins led to an explosion of tastes across southwest and Central Asia. Persian and Arabic chefs combined techniques, ideas, and ingredients, fueling a greater demand for new spices and foods. The court of the Umayyad Caliphate inspired this fusion by bringing in cooks from afar. The growing taste for spices and new flavors spurred global exploration; the flavors and scents of distant islands in the Indian Ocean and the mountains of China led to culinary assimilation and economic imperialism. During the Golden Age, Islamic chemists also experimented with alchemy, explorers mapped new lands in East Asia, politicians dabbled in democracy, and cooks mused over new ingredients and tastes that survive in modern kitchens.

During the early centuries of the Abbasid Caliphate, trade flourished along the Silk Road. Farsi-speaking immigrants developed alliances with Persian and Sogdian merchants to meet the great hunger of the Tang Dynasty for exotic and luxury goods. The Sogdian people, already dispersed across Central Asia, moved into what is now western China.[49] The Sogdians dominated trade from the fifth through the eighth century, centering on Panjikent, Samarkand, and Bukhara.[50] They were only one of the ethnic groups moving goods into and out of China. During this period, all of Central Asia saw a boom in the arts and cultural enterprises, including professional singers and dancers, who were trained from youth in guilds to perform in Chinese and Arab courts. Ultimately, the flourishing of cuisines and the arts was felt not just by the Umayyad political elite, but by all of Central Asia.

Another invading force, the Mongols in the early thirteenth century, transformed the Silk Road once again. In China, the Song Dynasty (960–1279) had nurtured a wealthy and vibrant trade network that connected Asia to Europe by both overland and nautical routes. By the 1210s, Genghis Khan had unified many of the tribes of the Mongolian steppe and set up his capital in Karakorum, starting the wheels turning that would eventually lead to the collapse of the Song and begin to erode trade along the northern routes of the Silk Road.

One of the confidants of Genghis Khan, Yelü Chucai, who accompanied the Mongol army on its westward campaign into Central Asia between 1218 and 1224, wrote a book praising the Western Regions. He pays tribute to Samarkand in a series of poems, noting that it was the most beautiful city in Central Asia, located on fertile land. "It was surrounded by numerous gardens. Every household had a garden, and all the gardens were well designed, with canals and water fountains that supplied water to round or square ponds. The landscape included rows of willows and cypress trees, and peach and plum orchards were shoulder to shoulder." He goes on to praise the fruits and wine of the city, stating that "in the dry summer, water was lifted from the river for irrigation. The grape wine was excellent."[51]

While the Mongol khans' empire was short-lived, another Turkic military leader would soon unite Inner Asia. The rise in power of Timur in Samarkand during the late 1300s and increasing commercial trade on the sea routes from China once again completely altered the face of trans-Eurasian trade. Different factors led to the slow decline of the northern routes of the Silk Road. Although the routes survived even after Russian imperial conquests, they never again saw the volume of transport that they did during their golden age. The transmission of intellectual culture, technology, raw and crafted materials, and human DNA through the mountains, valleys, and deserts of the Silk Road shaped European and Asian history. The spread of seeds and agricultural knowledge not only led to the adoption of farming in Central Eurasia but altered ecosystems and cuisines around the globe. The introduction of grain crops and crop rotation in East Asia and Europe (discussed later in this book) provided the nutritional surplus necessary to support growing populations and empires.

Having roughly laid out the historical and geographic trajectories of the Silk Road, I turn in the following chapters to an examination of the foods that were transported along it in different directions: their places of origin, what we know of their domestication, their spread into new regions along the Silk Road routes, and their subsequent influence on cuisines and cultures.

PART II

Artifacts of the Silk Road in Your Kitchen

CHAPTER FOUR

The Millets

Millets have fed far more of the world than the familiar, large-grained cereals such as wheat. The vernacular term *millet* has a variety of meanings, but most often it refers to any small-seeded cereal grain. These grains have been domesticated numerous times independently by prehistoric farmers the world over. In the southern-central states of North America (from Arkansas to Texas and up into Missouri) as long as four millennia ago, cultivators were growing may grass (*Phalaris caroliniana*). Two millennia ago in the Southwest, people were collecting and likely maintaining stands of Sonora dropseed grass (*Sporobolus* spp.). Sorghum, an African domesticate that is grown around the world today, is also often clumped into the millet category. A number of other millets originated in Africa, such as pearl millet (*Pennisetum glaucum*), finger millet (*Eleusine coracana*), teff (*Eragrostis tef*), and fonio (*Digitaria* spp.). A far wider array of millets has been grown by farmers in East Asia, including broomcorn, foxtail, and the barnyard millets (*Echinochloa crus-galli*), as well as little millet (*Panicum sumatrense*), kodo millet (*Paspalum scrobiculatum*), and browntop millet (*Urochloa* spp. and *Brachiaria* spp.). In this chapter, I discuss the varieties of this vital crop that have traditionally been grown in Central Asia.

BROOMCORN MILLET: AN EAST ASIAN CROP IN PRECLASSICAL EUROPE

Although broomcorn millet is known mainly as bird food in the industrialized world, it was one of the most important grains of the ancient world, sustaining the laborers, soldiers, and farmers who built Europe between the first millennium BC and the early second millennium AD. Broomcorn millet is the small, spherical, shiny seed found in most bird food mixes—not to be confused with the slightly larger, often reddish seed, which is sorghum. The crop is now enjoying a modest revival as an ingredient in the multigrain breads that have become trendy among wealthy American whole-food enthusiasts.

A number of ancient texts provide glimpses into the perceptions and uses of plants in southern Europe and southwest Asia. They mention harvesting and preparation methods, cultivation practices, and myths.

Herodotus (ca. 484–425 BC), whom Cicero referred to as the father of history, wrote extensively on the peoples to the north of Greece. Any scholar grappling with his *Histories* has to take every word with a grain of salt: like most Classical scholars, he made no attempt to distinguish myth from reality. He described the people he called the Scythians as cannabis-smoking, bloodthirsty mounted warriors who drank wine out of the skulls of their enemies. However, he also noted that Scythian farmers grew millet as one of their main crops, along with onions (which grow wild across the steppe). Refering to southwest Asia, also he stated: "The blades of the wheat and barley there are easily four fingers broad; and for millet and sesame, I will not say to what height they grow, though it is known to me; for I am well aware that even what I have said regarding grain is wholly disbelieved by those who have never visited Babylonia."[1]

Archaeologists have argued that one of the major driving mechanisms for the spread of both broomcorn and foxtail millet across Central and southwest Asia was the development of centralized state-run irrigation projects.[2] Herodotus comments on the expansive irrigation systems of southwest

Asia. These allowed farmers to grow crops year-round and encouraged crop-rotation cycles. Millet, which had once been a low-investment crop for the poor, became a significant crop in these rotation cycles, increasing its importance in Europe and southwest Asia.

People from the Balkan region had been cultivating broomcorn millet for almost two millennia and were likely some of the first Europeans to grow the crop, which they may have obtained from Central Asian pastoralists. In his Third Philippic, an oration given in 341 BC, Demosthenes warned the Athenian people that Philip was besieging Thrace through the winter "for the sake of the rye and millet of the Thracian store-pits." In his grand oration, attempting to rally the Greek citizens against the invasion of the "pastoralist" troops, Demosthenes focuses on the millet stores.[3]

Natural philosophers and physicians of the classical period also refer to varieties of millet. Hippocrates (460–370 BC) mentions roasted and fermented millet in his *De diaeta in morbis acutis*. Theophrastus (ca. 370–288/285 BC) described broomcorn millet in book 8, "Cereals and Legumes," of his *Historia plantarum*, and he mentions both of the small East Asian grains throughout his other writings. Born on the island of Lesbos, Theophrastus studied under Plato and was a close colleague of Aristotle. His great work represents one of the oldest botanical floras of Europe, which earned him the title of "the father of botany."[4]

Pedanios Dioscorides (ca. AD 40–90) was a Roman military doctor during the reign of Nero (AD 37–68). With the Roman army, he traveled across the empire, from Gaul to Asia Minor. By observing and recording details of the plants he encountered and their uses, he became essentially the world's first ethnobotanist. In his famous work *De materia medica*, written around AD 64, he describes roughly six hundred plants and plant products, a hundred more than Theophrastus had listed. Dioscorides discussed broomcorn (*Panicum miliaceum*) and foxtail (*Setaria italica*) millet, focusing on their health-giving qualities and praising a fermented beverage produced from the grain.[5]

The compilations of both Theophrastus and Dioscorides were dwarfed by the work of Pliny the Elder (AD 23–79), who mentioned nearly a thousand plants in his masterpiece *Naturalis historia*. Pliny is probably the most famous of the classical Greek biologists, and his work became one of the most widely referenced treaties on the natural world. His passion for knowledge led to his death while observing the eruption of Vesuvius, the volcano that destroyed Herculaneum and Pompeii. Archaeobotanical investigations in Pompeii illustrate the importance of millet in the diet of the city's residents.[6]

In book 18, chapter 10 of the *Naturalis historia*, titled "The Natural History of Grain," Pliny notes that both broomcorn and foxtail millets are summer-sown crops across the Mediterranean and in Anatolia, and later he claims that the grains can ripen in forty days. He observes, "There are several varieties of panic [foxtail millet], the mammose, for instance, the ears of which are in clusters with small edgings of down, the head of the plant being double; it is distinguished also according to the colour, the white, for instance, the black, the red, and the purple even. Several kinds of bread are made from millet, but very little from panic: there is no grain known that weighs heavier than millet, and which swells more in baking." In chapter 24, he lists several regions where farmers regularly grow millet, including Central Asia, where he claims it is the principal crop among the Scythians. "The Sarmatians [an Iranian empire in southwest Asia] live principally on this porridge, and even the raw meat, with the sole addition of mares' milk, or else blood extracted from the thigh of the horse." Pliny reiterates in the following book that both millets are drought tolerant and that they can be sown in the summer on poor soils; he notes that millet is a major crop in the Pontic steppe region south of the Black Sea. In a discussion of what he calls "artificial wines," he provides recipes for millet beer.[7] Although these grains (like rice) would have been consumed mainly in the form of porridge in their East Asian lands of origin, in the Central Asian culinary world they were also transformed into unleavened bread and beer.

Strabo (64/63 BC—ca. AD 24), in his *Geographica*, describes peoples and their customs across the Classical world; when discussing the Iapydes, a population in modern-day Croatia, he notes that "their lands are poor, the people living for the most part on spelt and millet. Their armor is Celtic, and they are tattooed like the rest of the Illyrians and the Miracians." He further notes that millet was cultivated in Gaul, in regions where grapes and other fruits were difficult to grow. Strabo also claims that on the banks of the Po River in northern Italy, the crop-rotation cycle includes millets: "They say that some of the plains are cropped all the year round; twice with rye, the third time with panic, and occasionally a fourth time with vegetables." Later, in discussing the people of the Pontic Steppe region, Strabo specifically notes the ancient towns of Pontic Comana, Dazimonitis, and Gaziura and states that the most important crop in this region is millet, both broomcorn and foxtail.[8]

Other Greek and Roman authors who mention millet include Hesiod (eighth century BC), in his *Shield of Heracles*, where he refers to millets as a summer-sown crop.[9] Xenophon (430–354 BC) mentions both foxtail and broomcorn millet among the crops grown in ancient Cilicia, a Persian region in southern Anatolia.[10] Polybius (ca. 200–118 BC), in a chapter of his *Histories* titled "Grain Production in Cisalpine Gaul," observes that copious amounts of both broomcorn and foxtail millet were grown in the Cisalpine Gaul regions, now the northern Italian Alps.[11] References to these crops also appear in the writings of Aristotle (371–ca. 287 BC); Cato the Elder (234–149 BC), in his *De agri cultura;* Varro (116–28 BC), in his *De re rustica* (37 BC); Lucius Junius Moderatus Columella (AD 4–70), in his own *De re rustica;* and Virgil (70–19 BC), in the *Georgics*. *Apicius*, a collection of recipes from Classical Rome, provides several recipes using millet.[12]

These Classical texts were directed toward an affluent and educated audience, not the plebeians who were toiling in the fields.[13] The villas of Roman citizens and Greek landowners were commonly planted with elite crops, notably grapes, olives, and fruits; hence, these texts are biased away from discussions of grains, especially grains considered to be the food of the poor. Therefore, millets were likely much more prominent than we might infer from these texts alone. The textual evidence attesting to the use of both types of millet in Europe in the first millennium BC is supported by archaeobotanical discoveries. The importance of broomcorn millet in the Roman world is illustrated by an archaeobotanical study at Pompeii.[14] An examination of botanical remains from numerous households in regio VI, insula I of the city revealed that broomcorn millet was especially common in these households, over three times more common than barley.

How broomcorn millet spread to Eastern Europe and Inner Asia by the first millennium BC and became such a significant crop has long been a mystery and a topic of debate. The early spread of crops through the mountain valleys of Central Asia played a fascinating role in establishing this crop across the Old World.

The Origins of Broomcorn Millet

MONOPHYLY OR POLYPHYLY?

Until recently many questions remained unanswered regarding the history of broomcorn millet, the grain that fed billions of peasants, small-scale farmers, and pastoralists across Eurasia for millennia: where and when the grain was domesticated, whether it was domesticated once or multiple times, and how it made its way across two continents before any written record of it existed. Furthermore, the wild progenitor, or parent population, for *P. miliaceum* has never been located; the question has typically been evaded by suggesting that this mysterious *Panicum* must exist (or have existed) somewhere in Central Eurasia.[15]

The most interesting chapter in the narrative of this millet is the debate over whether it had one or two origins: in botanical terms, monophyly versus polyphyly. Many aspects of this debate first arose in the Caucasus Mountains of

Georgia in the mid-1970s. Gorislava N. Lisitsyna (or Lisitsina) was the resident paleobotanist at the Institute of Archaeology of the Academy of Sciences in Moscow during much of the 1960s and 1970s, one of the few archaeobotanical specialists working in the former Soviet Union. She studied early agriculture in southern Turkmenistan (then part of the Soviet Union), specifically irrigation systems. By the mid-1970s she had turned her focus toward the Caucasus.

In 1977, Lisitsyna and her colleague L. V. Prishchepenko published a summary of archaeobotanical studies they had conducted at the sites of Kjultepe, Arukhlo, Imirisgora, and Chokh in Georgia and Azerbaijan.[16] It was already well accepted that agriculture in this part of the world stretched deep into the human past. Based on material recovered from these sites, the Soviet paleobotanists argued that there was a productive agricultural economy in the region by the fifth millennium BC. Among the finds were preserved grains of broomcorn millet. These probably came as no surprise to Lisitsyna, because many Russians have grown up eating various forms of kasha, or porridge, for breakfast, a common one being millet porridge. In a 1984 book chapter, Lisitsyna used these data to assert that there was a Vavilovian center of domestication within the political boundaries of the Soviet Union.[17] Nicolai Ivanovich Vavilov is often praised as the greatest of all Soviet botanists (the more so after his martyrdom in an early Stalinist purge during World War II), and Lisitsyna was no doubt strongly influenced by his work on the origins of cultivated plants. In the following decades, many scholars would build on her idea. During the Soviet period, it was common to study early agricultural grains by examining the impressions of those grains left on fired clay vessels. Currently there are thirty-one recorded sites across Eastern Europe with reports of broomcorn millet predating 5000 BC, and far more sites with less clear accounts of impressions of millet or undifferentiated grains of millet, *Setaria* sp., or simply *Panicum* sp.[18]

Evidence pointing to the domestication of broomcorn millet in Eastern Europe is countered by the fact that both foxtail and broomcorn millet have a long history in East Asian cuisine (where they are used for much more than just a breakfast gruel). They can be traced back several millennia in Chinese sources. In 1980, at the archaeological site of Cishan (6100–5600 BC) in the Taihang Mountains of northern China, a large cache deposit of grains was excavated, supplying evidence of broomcorn millet cultivation predating the finds in the Caucasus.[19]

Data attesting to the cultivation of this grain on the opposite sides of two separate continents more than seven thousand years ago cannot be easily explained. There is no other evidence for contact between these cultures, and no other evidence of grains spreading so far this early. There are two possible scenarios: that broomcorn millet was domesticated twice—in northeastern China and in the Caucasus or elsewhere in Eurasia (a hypothesis favored by the famous scholar of the "origins of agriculture" debate, Jack Harlan);[20] or that one crop crossed two continents without any other crops or clear evidence of material culture accompanying it and before any other evidence for contact along the Silk Road.

The single-versus-double-origins debate was resurrected in 2011 with the sequencing of the *P. miliaceum* genome.[21] That same year, a team of archaeologists and biologists from Cambridge University published a genetics-based population study, using genetic primers first identified in a larger genomics study. This 2011 study looks at the distribution of specific genes (alleles) in landrace varieties of millet grown across Eurasia. The researchers identified two genetically distinct populations of broomcorn millet—one in Eastern Europe and one in East Asia—that have clearly been reproductively isolated for a considerable period. As the authors point out, this genetic isolation could have resulted either from two independent domestication events or from an early spread of a small population of the cultivated plant into Eastern Europe and its subsequent isolation from the parent populations.[22] The latter possibility could theoretically be explained by a long-distance traveler who carried a small sack of grain and started a new field of millet in the Caucasus, ten thousand kilometers from home. Hence either of the earlier proposed scenarios could explain the results of the genetic study.

Revisiting the dozens of published accounts of millet finds in Europe that predate 5000 BC, scholars have started questioning some of the claims.[23] They point out that in relation to the extensive archaeobotanical analyses that have been conducted at sites in Europe, there have been only a handful of finds of early millet.[24] Furthermore, these finds often involve only a few grains, as opposed to finds of thousands of grains of wheat (*Triticum* sp.) and barley (*Hordeum vulgare*) at the same sites. They suggest that at least some of these millet finds could represent particularly large specimens of wild panicoid grasses, which may grow as weeds in agricultural fields. Giedre Motuzaite-Matuzeviciute and her colleagues from Cambridge point out that because these finds were reported as long as fifty years ago, in many cases the excavated grains or impressions no longer exist, making it impossible to check the original identifications or to redate the finds. However, her research team did track down ten key specimens that they consider representative of the majority of the claims and submitted them for radiocarbon dating. Despite the seemingly massive evidence for early millet cultivation in Europe, Motuzaite-Matuzeviciute and colleagues state that "the dates indicate that the chronology previously proposed for the substantial number of Central and Eastern European broomcorn millet macrofossils was too early by at least 3500 years."[25]

This article is unprecedented in the field of archaeobotany; it suggests that the dozens of published accounts of these finds are unreliable, and it directly refutes numerous published site reports. Furthermore, it paints a completely new picture of early agriculture in Eurasia and changes how we look at the spread of crops along the proto-Silk Road.

In the end, it appears that neither of the two scenarios—polyphyly, with millet originating in both East Asia and Eastern Europe, or monophyly, with an early division and isolation of a single population—holds up. Assuming that the conclusions of Motuzaite-Matuzeviciute and her colleagues, including the archaeogenticist Harriet Hunt, are correct, broomcorn millet did not even make it into Europe until the early second millennium BC. The finds of earlier millet grains in Europe must represent either misidentifications or intrusive grains from more recent archaeological sediment levels.

MILLET'S HOMELAND IN NORTHEASTERN CHINA

Both broomcorn and foxtail millet were domesticated in roughly the same geographic region and prehistoric period, which is interesting given that they are often paired in cultivation systems across the Old World but did not spread out of East Asia together. Both crops seem to originate somewhere in the vicinity of the Yellow River in northern China.[26] Their tolerance for drought may be an adaptive response to the conditions in the semiarid grasslands or open meadows in the hills north of the Yellow River.[27] The earliest finds of preserved grains of carbonized broomcorn millet come from the Dadiwan site (ca. 5900 BC), northeastern China, although there are finds from several roughly contemporaneous sites in the same general region (map 2).[28] For example, large quantities of carbonized grains have been found at the site of Cishan, although their provenance and dating have been controversial. The widespread cultivation of broomcorn millet across northeastern China by the sixth millennium BC is further verified by finds from the site of Xinglonggou in Inner Mongolia (5670–5610 BC). Over 1,400 charred broomcorn millet grains (and 60 foxtail millet grains) were recovered from Xinglonggou; these grains were directly dated and verified by three independent laboratories.[29]

Stable carbon isotope research (δ13C) over the past few years has significantly contributed to our understanding of the spread of millets. Because millets use a different system of photosynthesis (the C_4 system) than most other plants, they have a different ratio of the carbon isotopes that are used for archaeological understanding of human diet. The signature of this ratio is visible in the bones of the ancient humans who ate the millet. This kind of

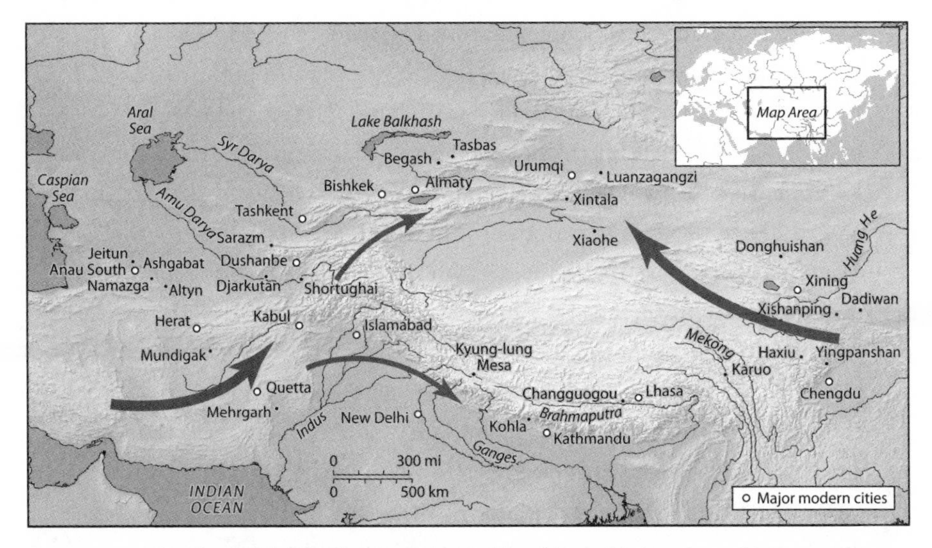

Map 2. Key archaeological sites in the mountain corridor of Central Asia and possible routes of crop diffusion. Crops spread through the rich, arable mountain valleys, which had higher densities of human occupation, glacial-melt streams, and rain-fed meadows.

research at Dadiwan has further supported these dates and thus the idea that by this time millet cultivation was widespread in this area.[30]

Complementing the macrobotanical conclusions, the low level of δ13C enrichment found in remains of human bones excavated at Dadiwan seems to indicate that between around 5900 and 5200 BC, the diet of people at the site included limited quantities of millet. These data likely illustrate that an early form of low-investment millet cultivation was practiced and that there was a mixed economy that still relied heavily on hunting and gathering. However, human bones from the site dated later than 3900 BC show enriched levels of carbon isotopes, likely a result of intensified millet cultivation.[31] Other scholars have noted, based on the archaeological record, that the period 5000–3000 BC was also a period of rapid cultural development, clearly a response to the intensification of agriculture and a grain surplus.[32]

Archaeobotanists working in many (though not all) of the world's centers of plant domestication can trace the long process, from the foraging of the wild relatives of our modern food plants through the gradual process of morphological change into plant varieties we recognize today. In southwest Asia, there is copious evidence of humans collecting wild wheats and barley and evidence for the gradual process of domestication in the southwest Asian founder crops. While the macrobotanical data suggest that millet cultivation began in the sixth millennium BC, they have not provided conclusive evidence for this kind of predomestication cultivation or for the interaction of people with these crops during the domestication process. This lack of evidence for the millets has left a lot of room for speculation about possible earlier dates of domestication. As with many of the other major world grains, such as rice and maize, microbotanical data have provided highly controversial dates for the earliest millet domestication that predate the macrobotanical evidence by several millennia.

The use of microscopic remains of plants to identify the earliest agriculture has focused on phytoliths, which are small siliceous particles inside and between cell walls; starch grains, which are the packed glucose-storage particles inside grains and roots; and pollen. While these methods are advancing scholarship, the reliability of identification and dating methods is a matter of debate.

One month after the publication of the stable carbon isotope data from the Dadiwan skeletons, which supports the current macrobotanical conclusions, another team of scholars published another article in the same journal boldly claiming that they had evidence of millet "domestication" at the Cishan site, further sensationalizing the already controversial Cishan remains.[33] This claim, should it hold up to scientific scrutiny, pushes the date for the earliest agriculture in East Asia back almost to the period of the earliest known plant cultivation in the world, in southwest Asia. This claim has excited and elicited support from many China historians, scholars, and archaeologists.[34] The authors base the claim that they have a possible fore-runner for the global origins of agriculture on criteria that they developed and published in an article just one month prior to the publication of their surprising conclusions.[35] These new criteria are intended to differentiate not only between phytoliths of domesticated foxtail and broomcorn millet but also between wild and domesticated lines of the two crops. The authors claim, based on qualitative observations of phytoliths from a single wild *Setaria viridis* specimen and two wild *Panicum bisulcatum* specimens, that "a species-specific identification of phytoliths is possible for *S. italica* and *P. miliaceum* because they have typically well-defined silica skeletons that are distinguishable from those in *P. bisulcatum*, *S. viridis*, and *S. plicata*, which have no such demonstrable patterns, [but] additional studies are needed to confirm the observations."[36]

Other aspects of this study raise further questions about the researchers' dating of the Cishan site. They state, for example, that grain crops were recovered in situ from eighty-eight storage pits at the site. They note that the grains were "well-preserved" but then say that they were very poorly preserved, having "been oxidized to ashes soon after they were exposed to air."[37] Relying solely on phytoliths leaves a lot of room for future debate, although a team of Chinese scholars using starch grain analysis, another microbotanical approach, at a nearby, culturally related site actually argues for an even earlier domestication of millet.[38] However, these starch grain data have also raised red flags in the scientific community. Many scholars working on domestication in East Asia have pointed out that both lines of microbotanical data are problematic and require further investigation.[39]

This starch grain study of early millet domestication in China is one of the few examples of a microbotanical study being tested by another team running an analysis on material from the same archaeological site. Starch grains can survive in specific conditions for millennia: archaeobotanists often recover them from crevasses on stone grinding tools or ceramic vessels.

A close analysis of a set of features on the grains can help determine the general groups of plants that produced them. In this case, both teams of researchers were looking for starch grains deposited on grinding stones from the site of Donghulin, in northeastern China, dating to 9000–7500 BC. Interestingly, they came up with very different results.[40] The first study concluded that the starch grains recovered from Donghulin were from acorns (*Quercus*, *Lithocarpus*, or *Cyclobalanopsis*) and noted that the starch grains did not match the millets or grasses in size or general morphology.[41] The idea that people in this region of China at this time period relied heavily on acorns (as opposed to grass seeds) is supported by a growing number of macrobotanical studies and other starch grain studies.[42] The second team looked at stored grinding stones from the same site and presumably from the same excavations, arguing with confidence that the starch grains were from both domesticated broomcorn and foxtail millet. They claim that their "data extend the record of millet use in China by nearly 1,000 y[ears] [a millennium earlier than even the authors of the phytolith study were arguing for], and the record of foxtail millet in the region by at least two millennia."[43] They dismiss the first starch grain study by referring to the identification of acorn starch as "tentative."

While there are numerous other data sets supporting acorn collecting in this period and region, there is no other evidence for millet domestication. Furthermore, this study raises more questions than it resolves: for example, the researchers claim that they also retrieved millet starch grains from carbonized residue on a sherd from a cooked ceramic vessel. But because grains are highly sensitive to heat and degrade when cooked, a find of cooked and carbonized starch grains on a vessel would be unexpected. Nor does it seem likely that the occupants of the site would grind millet into flour and then boil it in a pot, which would presumably yield a paste or gruel rather than whole-grain porridge. Furthermore, although the authors themselves note that many studies of other grain crops have shown that the domestication process took two thousand to three thousand years, they note that all of the earliest microscopic remains of these plants found were from domesticated (not wild) grains. Their discovery would thus seem to push the earliest cultivation of these crops into the Pleistocene. Beyond those issues, it simply seems hard to believe that both crops were fully domesticated by 9000 BC, as the authors assert.

Further illustrating the prominence of these debates, another recent publication in the same scholarly journal as the previous studies—also on ancient starch and phytoliths—claims to have identified barley phytoliths on vessels from the site of Mijiaya in Shaanxi in northern China that they claim are five thousand years old. Their claim dates the cultivation of barley to roughly a millennium before any other evidence for the grain in East Asia—a conclusion that seems to stand in clear contradiction to the rapidly growing number of macrobotanical studies in China. They note that their "recently developed method based on phytolith morphometrics [a set of specific measurements of the shape]" allows them to identify remains of barley that predate "macrobotanical remains of barley by 1,000 y[ears]."[44]

With the rapidly increasing rate of archaeological investigation and rescue excavations in China, because of an unprecedented rate of urban sprawl, the earliest date and location of millet cultivation will surely be determined during the next few years. However, regardless of whether we choose to believe the macrobotanical data (placing millet domestication around 5900 BC), the phytolith data (8300–6700 BC), or the starch grain analysis (9000–7500 BC), it is clear that both crops originated in northeastern China and have had long relationships with humanity.

Broomcorn Millet in Central Asia

If we accept the conclusion that millet did not reach Central Asia until the end of the third millennium BC and Eastern Europe some time later, then its movement out of East Asia would coincide with the earliest iterations of the

Silk Road (or what has been called the Inner Asian Mountain Corridor).[45] Broomcorn millet and highly compact varieties of free-threshing wheat and barley were the earliest crops to move through Central Asia.[46] Well adapted to rain-fed agriculture in the mountain foothills, they spread along the Tien Shan Mountains and the edge of the Tibetan Plateau, then to southern Central Asia and the Indus Valley, and eventually to Anatolia and the Balkans.[47] From there millet quickly spread into Europe, making its way into the western Caucasus by the middle of the second millennium BC.[48]

This initial westward spread of broomcorn millet took place a millennium before the inception of large-scale public works irrigation projects. Because of millet's relative ease of cultivation, its spread seems to have been associated with pastoralists or low-investment (nonirrigated), small-scale agriculture.[49] The foothills of Central Asia were well suited to dry farming or limited gravity-fed irrigation from glacial-melt streams and provided a suitable setting for pastoralists to plant small plots of low-investment millet. These plantings enabled herders or small-scale farmers to insure themselves against the potentially catastrophic consequences of a bad harvest of other crops or loss of livestock. Interestingly, broomcorn millet is still associated with mobile peoples of Inner Asia today.[50]

The two East Asian millets are highly tolerant to ecological stressors. Strabo observes in his *Geographica* that "millet is the greatest preventive of famine, since it withstands every unfavourable weather, and can never fail, even though there be scarcity of every other grain."[51] Scholars have identified three traits that would have made the millets especially well suited for integration into a mobile pastoral economy. Because the plants are drought tolerant, they do not require the development of large-scale irrigation systems: they can grow in small plots near streams or in water catchments. Because they have a high yield per plant, herders could carry a small sack of grains for sowing. Finally, they have a short growing season, a summer crop could be sown and harvested before the seasonal migration to winter camps.[52]

In addition, broomcorn millet's shallow root system makes plowing unnecessary.[53] Probably because of its rapid growth (taking as little as sixty days from planting to harvest, given optimal conditions), it requires only about half as much water as free-threshing wheat.[54] Another potentially appealing characteristic of millet is its small size, which makes it quicker to cook and thus requires less fuel.[55] This would have been particularly important in a landscape with dwindling wood reserves, where the main fuel source was herd-animal dung—a great sustainable fuel, but one that produces less heat than wood.

The accounts of early European explorers in Central Asia are full of mentions of low-investment cultivation. Before Russian influence, many mobile peoples in Central Asia were growing broomcorn millet and lesser quantities of barley in small, low-elevation fields.[56] These plots were usually within five kilometers of a fall or spring camp but as much as thirty or forty kilometers from a winter camp. Because these crops needed so little tending, the herders had to ride out to the plots only a few times during the growing season. Fields were visited for planting in April and harvesting in October, and little attention and no irrigation were required in between. The plots used for cultivation were relatively small, rarely larger than 1.5–2 hectares (3.7–5 acres). The arid steppe environment is not suitable for most crops without extensive irrigation, so the plots were often in river valleys or near a water source.[57]

Rona-Tas's 1959 study of agricultural practices among mobile pastoralists in the Selenga River valley of western Mongolia, as discussed by Nicola di Cosmo, provides a good summary of low-investment agriculture among Central Eurasian pastoralists. Rona-Tas observed small plots near river banks being turned using wooden plows; soil clods were broken up by hand, and wheat, barley, or rye (*Secale cereal*) seeds were hand planted. The herders then took their herds to summer pastures and did not return until autumn. Rona-Tas also noted that harvesting was done by hand without the aid of a sickle. Winnowing was done with large wooden shovels, and grain was ground with a horse-driven mill. These ephemeral tools, if used by prehistoric populations in Central Asia, would not have left archaeological evidence.[58]

Similar ethnohistorical accounts of small-scale, low-investment farming come from throughout the mountainous and desert regions of Central Eurasia. The geographer, historian, and Silk Road explorer Owen Lattimore insisted that steppe populations were self-sufficient for food.[59] Similarly, Khalel Argynbaev notes that "at the start of the century, dry farming in the Semirech'ye province was introduced only under conditions of small plots, scattered throughout mountain fields."[60] However, there is also evidence of irrigated agriculture along major rivers through the medieval period and earlier, as evidenced by large towns and settlements.[61] Early agriculture was practiced near rivers and springs, utilizing mountain rainfall and glacial melt; dry farming was practiced in higher elevations and foothills.[62]

Other historians have echoed Herodotus in noting the importance of the millets among mobile steppe populations, specifically the Scythians (although the identity of this group remains uncertain).[63] Pliny recorded millet cultivation to the north of the empire, in the steppe regions. Columella (AD 4–70), in his treatises on agriculture, the most important works on the topic from the Roman period, observes that "in many countries the peasants subsist on food made from [panic and millet]."[64]

Classical texts are not the only historical references to millet cultivation in Central Asia. In the Geographical Part of the *Nuzhat-al-Qulub*, composed by Hamd-Allah Mustawfi of Qazwin in 1340, Turkic lands, peoples, and cultures are described, often along with the main crops grown in those regions. Interestingly, the peoples of the Golden Horde (thirteenth century) are described as being pastoralists who also grew millet and summer wheat. He stated that "the climate here is extremely cold; and they raise but scanty crops, growing for the most part millet and summer corn [cereal], but neither cotton, grapes nor any other fruits can be brought to ripen. These folk have, however, much cattle, seeing that for their food and means of living they depend on cattle breeding. Further the produce of the rock-crystal mines here is very considerable." He goes on in the same chapter to state: "The crops consist of wheat in small quantity, but millet and other summer cereals are good both in quality and quantity. Grapes, water-melons and other fruits are here remarkably rare, and no cotton is grown. The pasturage, however, being excellent, horses and cattle are numerous, and the population for the most part subsists on the produce thereof."[65]

While Arab sources do not discuss the preparation of millet in as much detail as other plant-based foods (perhaps because it was considered a low-status crop), a few early Islamic sources from southwest Asia note that thin millet breads were produced in clay tannour or tandoor ovens, with the millet sometimes mixed with lentil flour. The accounts of early travelers through modern-day Syria and Iran also mention millet breads.[66] The twelfth-century *Book of Nabatean Agriculture* compared the preparation of millet and rice breads and noted that millet was cultivated in the summer, like rice, but with less water.[67] Ibn al-Awwam noted that there were some areas where it was possible to cultivate millet in the summer without irrigation.[68] El-Samarraie (summarized by Delwen Samuel) discusses ninth-century sources from Iraq that mention mixing millet and barley flour together to produce a dense bread that some ancient writers, such as Ibn Wahshiyya, praised as suitable for men doing physical labor.[69]

Broomcorn millet also offered a reliable alternative to the labor-intensive cereal grains in the agricultural centers of Europe and Western Asia. It became culturally significant across much of Inner Asia; for example, at the midyear festival in the Zoroastrian tradition, on the second day of the month of Faghakan, followers eat a dessert made from millet, butter, milk, and sugar.[70]

There is a clear divide in the culinary treatment of grains between East and Central Asia. In East Asia, grains are boiled and steamed (and made into buns, noodles, rice, and porridge). In Central Asia and Europe, grains are more often ground and baked into bread.[71] When millet crossed this culinary boundary, it appears to have taken on a dual role, being used to make both a boiled porridge and various forms of bread. Several ancient texts by Dioscorides, Pliny, and Columella, among others, discuss ways to bake bread from millet flour or mixed flours.[72] These texts indicate that millet bread was commonly consumed but was probably a food for the plebeians or working class. Many other textual sources mention millet porridge, often consumed with milk.

The millets' short growing season, drought tolerance, and preference for warm conditions make them well suited for growing as a summer crop in a crop rotation, and accounts by Virgil and Columella, among others, corroborate this practice.[73] Even so, summer plantings would have required irrigation. Hence it was not until after large-scale irrigation projects were implemented across southern Europe and South Asia that the millets became an important crop.[74]

Archaeobotanical Data for Broomcorn Millet

The rapidly growing archaeobotanical evidence for broomcorn millet consumption in ancient Inner Asia allows us to trace the spread of this crop rather precisely.[75] Much of the earliest evidence for broomcorn millet outside East Asia comes from small-scale agropastoral homesteads, which many archaeologists interpret as mobile pastoral camps. The earliest grains so far recovered outside East Asia come from my examination of a human cremation cist, presumably burned and interred as a grave offering with the remains of the deceased at the site of Begash (2200 BC; see map 1).[76] Begash is located in eastern Kazakhstan, in northern Central Asia, and we can assume that millet spread via Xinjiang to reach Central Asia over only a few centuries.

Impressions of millet grains have also been reported from the large urban agricultural center of Gonur Depe on the Murghab Delta (2500–1700 BC; see map 1), but these grains may come from second-millennium BC occupation layers at the site.[77] However, radiocarbon dating of a millet grain that I recovered from the ancient city of Adji Kui, not far from the larger metropolitan center of Gonur, provided a calibrated date of 2272–1961 BC.[78] Adji Kui was inhabited through the third and second millennia BC and was an important trading stop on exchange routes in southern Central Asia. The excavators collected sediment samples from the city in 2013 and shipped them to me for analysis of the botanical remains. These excavators also noted the recovery of ceramics and material culture similar in style and form to artifacts from northern Central Asia. They argued that these suggested links between "nomadic" peoples and the city's inhabitants. These pastoral links may explain why millet has been recovered from pastoral camps and low-investment homesteads but not from urban centers dating to the same period.

I have recovered ancient carbonized broomcorn millet grains from small-scale encampment or homestead sites in southern Central Asia, such as Ojakly in Turkmenistan (ca. 1600 BC)[79] and the site of 1211 (ca. 1200 BC, also in Turkmenistan). I sorted through the archaeobotanical remains from both sites and personally collected botanical samples from Ojakly in 2011. These sediment samples (and other samples from archaeological sites across Central Asia) were processed for botanical remains in the field using geological sieves and an archaeobotanical method for separating plant remains from sediment, which relies on buoyancy in water (flotation samples). During follow-up seasons at the sites in southern Turkmenistan, my American and Italian collaborators collected samples for me. One of these samples contained a small cache or collection of 247 broomcorn millet grains. All of these small occupation sites are located about twenty kilometers from Gonur Depe.[80]

The presence of millet at both the northern and southern ends of Central Asia by the late third or early second millennium BC suggests that the crop likely spread rapidly through the mountain foothills. By the first millennium BC, millet was an important part of the economy, as attested by finds of the crop in most archaeobotanical assemblages.

In addition, millet has been recovered from other southwest Asian archaeological sites, such as the early-first-millennium BC layers of Tahirbai

Depe and second-millennium BC occupation layers at Shortughai (level II, period I; see map 2).[81] Broomcorn millet grains have also been identified from deposits at the Dam Dam Cheshme rock shelter in Turkmenistan (1200–800 BC). Furthermore, broomcorn millet made its way into the Harappan culture of the Indus River Valley roughly three to five thousand years ago. Millet was recovered at one of the main Harappan archaeological sites, a village or city referred to as Pirak; the grains date to roughly 2000 BC.[82] Large quantities of broomcorn millet grains dating from 1900 to 1550 BC were recovered from even farther south, at Haftavan, Iran.[83]

By the first millennium BC, millet had been integrated into the increasingly complex farming systems of Central and West Asia. It was one of the abundant crops recovered from the small village site of Tuzusai (410–150 BC; see map 1).[84] Several crops and a variety of herd animals were maintained around the site. Additional data from ongoing research have illustrated that the entire foothill region was occupied by farming communities, which likely relied on crop rotations or else used millets as a risk-reducing fallback crop. Similarly, at the site of Tahirbaj Tepe, Turkmenistan (ca. 650–500 BC), large quantities of broomcorn millet grains were recovered, and at the roughly contemporary site of Kyzyl Tepa, Uzbekistan, both broomcorn and foxtail millet were identified.[85] Millet has been reported from the Neo-Assyrian sites of Nimrud (Ziggurat Terrace) and Fort Shalmaneser, both in Iraq (late seventh century BC), as well as Deir Alla, Jordan, Phase VI (650 BC).[86]

While foxtail millet appears to have trailed far behind broomcorn millet in spreading across the Old World, it had crossed South Asia by the first millennium BC. Both millets had dispersed across the Arabian Sea to Yemen and then to Sudan by the middle of the second millennium BC.[87] Broomcorn millet had also made its way into the Balkans and northern Greece by the beginning of the second millennium BC.[88] It was recovered from archaeobotanical samples at the ancient city of Troy, dating to roughly 1550 BC, as well as

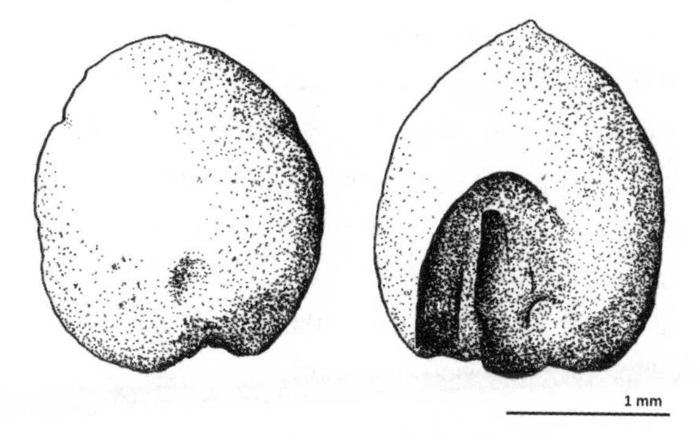

Figure 7. Dorsal and ventral views of a carbonized broomcorn millet grain from the ancient village of Tuzusai on the Talgar alluvial fan in the Semirech'ye region of southeastern Kazakhstan, dating to the second half of the first millennium BC.

other contemporary settlement sites in Turkey and the eastern Mediterranean.[89] The sporadic reports of early millet finds across southwest Asia leave open the question of how they made it to the Balkans and to North Africa.[90] Recently a deposit of millet grains articulated into a large burned clump, like overcooked porridge, was collected from the Guamsky Grot rock shelter in the western Caucasus. This find, dating to the end of the second millennium BC, confirms the presence of millet in Eastern Europe and illustrates how fast this grain traveled with early people.[91]

Carbon and nitrogen isotope studies of archaeological finds across North Asia have intensified over the past five years. In these studies, enriched levels of carbon in human bones are taken as evidence that the individual was consuming a lot of C_4 plants, which include the millets. Once the limitations of the data are acknowledged, stable carbon isotope analysis of human and animal bones provides supporting evidence to complement other studies. A study of 354 human and animal bones from thirty-seven archaeological sites (ca. 2700–1 BC) from across the Minusinsk Basin, Russia, found an interesting trend in the data.[92] The specialists noted a slight enrichment of the δ13C values from human bones after 1500 BC, indicating either an ecological or a dietary change. This would fit well with the archaeobotanically attested introduction of millet cultivation in the region.[93] It would not be a stretch, therefore, to argue that this transition illustrates a shift toward a mixed millet and pastoral economy in the Altai Mountains. The Minusinsk Basin isotope study is particularly important because of its large time scale, which provides isotope readings from humans before and after this transition.

Other follow-up studies consistently note high consumption of C_4 plants across Eurasia by 1500 BC. Another study in the Minusinsk Basin found enriched δ13C values in human remains from cemeteries at Ai-Dai, on the west bank of the Yenisey River (740–410 BC), and Aymyrlyg, in the Ulug-Khemski region of the Autonomous Republic of Tuva (mostly fifth through second centuries BC; see map 1).[94]

Another interesting isotope study, conducted on material from the Kostanai region of northern Kazakhstan, found no evidence for millet consumption—an important conclusion because it shows the temporal and spatial boundaries of the westward spread of this crop.[95] The team analyzed carbon isotopes in human bones from the cemeteries of Bestamak (2032–1640 BC), on the Buruktal tributary of the Ubagan River, and Lisakovsk (1860–1680 BC; see map 1), along the Tobol River. Following up on this study, a significantly broader analysis presented the results of a carbon isotope analysis of bone collagen from twenty-five archaeological sites, including 127 human and 109 animal bones, spanning all of Kazakhstan and the period from roughly 2920 BC to AD 1155.[96] The results support the argument that millet cultivation became a prevalent part of the economy in the foothills of Central Asia by the second millennium BC and eventually spread through southern Central Asia toward Europe.

Broomcorn Millet: Conclusions

Although it was gradually replaced by free-threshing wheat in Europe and rice in Asia, broomcorn millet was arguably the most influential crop of the ancient world. Its unique agronomic traits made it well suited for cultivation along the early routes of the Silk Road. The crop was originally domesticated in East Asia; from there it passed along the mountain foothills of Central Asia and into Europe by the second millennium BC. It was readily adopted by peasants, small-scale farmers, and pastoralists because its cultivation required relatively little labor. Among these communities, it reduced the risks associated with specialized economies and living on marginally productive land. Thus these grains, only about two millimeters in diameter, fed Silk Road porters and the laborers who built the empires of the Old World.

Millet gained further prominence in the latter half of the first millennium BC, with the development of centralized irrigation systems and collective labor. These systems allowed the introduction of summer irrigation, which enabled farmers to plant two (or sometimes three) rotations of crops on one field, as long as herd-animal dung was added to maintain soil nutrients.[97] With its tolerance for hot, dry conditions, millet enabled year-round crop cultivation, eliminating any period of rest for either the farmer or the soil. These crop-rotation cycles were highly productive and resulted in the great grain surpluses that supported imperial expansion. They enabled a larger portion of the population to divert their attention away from plowing soil and toward intellectual and artistic pursuits as well as political and military endeavors. Millet, therefore, became the crop that turned the chariot wheels of the Roman Empire and fed Eastern Europe. It nurtured Silk Road drovers and "nomads" across Asia; it filled the bellies of generations of Persian peasants and everyone in East Asia outside the rice-growing zones. Today, however, this grain, with its illustrious history, has been reduced to a children's breakfast food in Russia and bird food in Western Europe and America.

THE OTHER EAST ASIAN MILLET

Although foxtail millet is now a largely forgotten crop, its roots, like those of broomcorn millet, stretch deep into antiquity. Both are members of the Paniceae family of grasses. In many of the classical accounts, broomcorn (*Panicum miliaceum*), which is usually referred to as *milium* in ancient Latin sources, and foxtail (*Setaria italica*), which, confusingly for modern readers, is often referred to as *panic*, are discussed together. The two grains are often found together at archaeological sites across Eurasia dating from the first millennium BC, likely illustrating a practice of growing and cooking them together (figure 8). The two crops were traditionally grown as low-investment crops by Central Asians and on marginal soils by impoverished farmers across Eastern and southern Europe.

Foxtail millet was likely domesticated roughly contemporaneously with broomcorn millet on the open grasslands or the floodplains of the Yangtze River in northeastern China. Thus far, the oldest clear evidence for domesticated *Setaria*, from the Yuezhuang site (6000–5700 BC), roughly coincides with the onset of *Panicum* domestication.[98] However, for the millets of northeastern China, no evidence of the domestication process has been recovered, and there is no solid evidence for human gathering of wild grasses before domestication. This lack of a clear date or location for domestication in the millets has led to extensive debate.

Macrobotanical research at Shizitan Locality 9 (11,800–9,600 BC) in Shanxi Province in northeastern China has provided the only tentative evidence for human use of wild grasses before the domestication of the millets. Although the assemblage from the site does contain wild panicoid grass seed, it contains only eight seeds or seed fragments, including two *Echinochloa*, two *Setaria*, and a few from the plant families Poaceae and Amaranthaceae (also referred to as chenopods).[99] However, there is no reason to think that these few seeds represent economically significant plants, or that they are the result of human gathering of wild seeds. These happen to be the

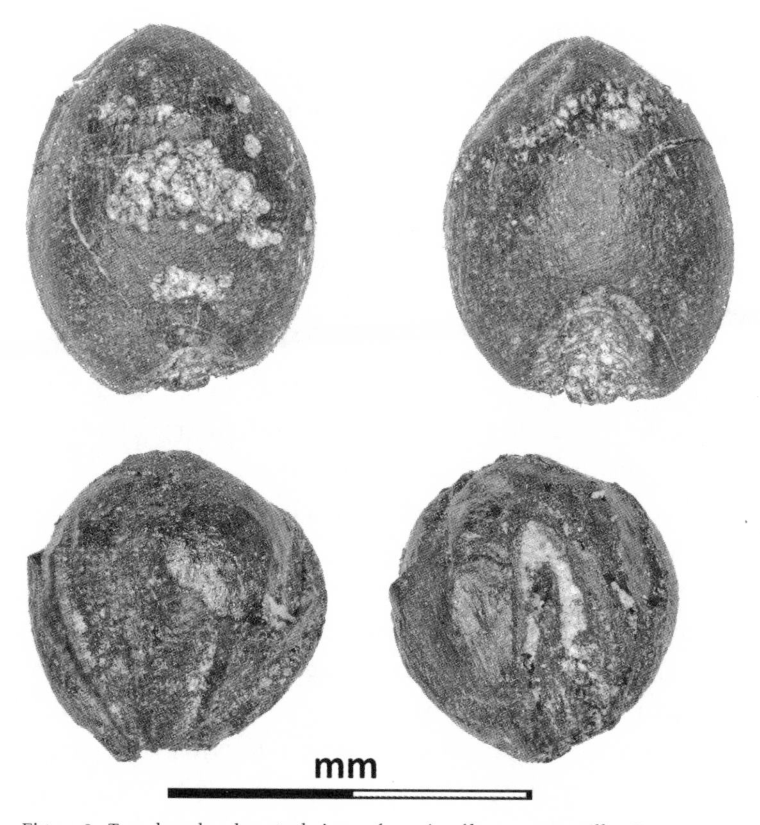

Figure 8. Top: dorsal and ventral views of a grain of broomcorn millet. Bottom: opposing views of a grain of foxtail millet, recovered during an archaeobotanical study at the Hellenistic site of Bash Tepa in Uzbekistan, excavated under the directorship of Sören Stark from the Institute for the Study of the Ancient World at New York University. Both grains are approximately two and a half millennia old.

most common categories of seeds identified in archaeobotanical assemblages from across northern Eurasia, and in many cases from the Northern Hemisphere. Many of the archaeobotanical assemblages from the grasslands of Asia are dominated by wild panicoids and chenopods, even from later periods when the gathering of wild seeds was unlikely.

The investigation of millet domestication is hindered by a lack of clear-cut morphological indicators for the early stages of domestication in macrobotanical remains. In many of the world's domesticated grains, especially those from the founder crops of southwest Asia (i.e., wheat and barley), the earliest phenotypical trait of domestication that archaeobotanists look for is a tough rachis, the small stem by which an individual grain or small cluster of grains is attached to the ear. In their wild form, most grains are programmed to detach easily after the grain ripens; however, in domesticated cereals, the grains remain attached to the ear throughout the harvesting process. This change is an inadvertent result of human harvesting with sickles: as people reap their harvest, the grains with a brittle rachis are dropped, and those with a tough rachis are collected, stored, and replanted for successive harvests. This change occurred in all our cereal crops, including wheat, barley, rye, oats (*Avena sativa*), and rice, and undoubtedly occurred in broomcorn and foxtail millet as well.

Detecting this change in millet, however, is difficult because the rachises are too small and fragile to survive in the archaeological record. Other early traits of domestication are also hard to detect, including reduced tillering, which refers to the number of stalks on a plant. Reducing tillering makes a grass plant look less bushy and more like a stalk. Additionally, the grains increased in overall size, and the number of grains on each ear (panicle) increased.[100] Increased grain size is a later development in foxtail millet, and early examples of the grain are hard to distinguish from its wild progenitor, *S. viridis*, or other wild *Setaria*. The ancient remains recovered from human occupation sites are hard to differentiate archaeobotanically, either between foxtail and broomcorn millet or between domesticated and wild lines of either type.[101]

Despite these uncertainties, it is clear that foxtail millet was well known across Eastern Europe by the later first millennium BC. As with broomcorn millet, the widespread incorporation of foxtail millet into the farming regimes of Eurasia was likely a result of centralized irrigation schemes.[102]

Foxtail millet was, however, likely known in Eurasia long before its general acceptance as a summer rotation crop. There are many reports of foxtail millet grains, and *Setaria* more generally, from archaeological sites that are supposed to date to more than four thousand years ago across Europe—although, as with the early reports of broomcorn millet, there are questions about their reliability.[103] One report of some reliability (my own), comes from the site of Tasbas (ca. 1400 BC) in the foothills of eastern Kazakhstan.[104] Only nine fragmentary foxtail millet grains were recovered from the site, compared to hundreds of other domesticated grains, such as free-threshing wheat, naked barley, and broomcorn millet. The fragmentary nature of the handful of foxtail grains makes their identification slightly problematic; however, it is supported by phytolith work.[105] These data suggest that the small-scale agro-pastoralists at Tasbas cultivated foxtail millet in the middle of the second millennium BC but that it was likely a crop of lower importance.

The range of early foxtail millet archaeobotanical remains in Central and South Asia confirms that it was widely grown by the first millennium BC. Both millets have been identified in the first-millennium BC layers at the ancient city of Gordion in central Anatolia.[106] The fortified Achaemenid city of Kyzyl Tepa, in the Surkhandarya region of Uzbekistan, dating from the mid-first millennium BC, has the best-preserved grains (including rachises) of domesticated foxtail millet found in Central Asia.[107] These abundant remains include more than 2,000 grains of broomcorn millet and 1,500 of foxtail.[108] Carbonized remains of both millets were also recovered from the ancient village of Tuzusai, dating to 410–150 BC.[109] In West Asia, large caches of foxtail millet, roughly fifteen liters in total, were found at Tille Hoyuk in southeastern Turkey, dating to sometime after 708 BC.[110]

The cultivation of millet was part of a pattern of greater agricultural production during the later first millennium BC. The use of centralized irrigation systems, elaborate crop-rotation cycles, and novel irrigation technologies, such as the development of qanats (underground channels), gave rise to a crop surplus that led to population increases, territorial expansion, and greater craft specialization. By the late first millennium AD, summer irrigation and crop rotations were commonplace across southwest and Central Asia.[111] These agricultural innovations also led to the cultivation of other new crops, including cotton and an ever-increasing variety of fruit trees.

CHAPTER FIVE

Rice and Other Ancient Grains

RICE: A TALE OF TWO GRAINS

Two foundational concepts in Chinese cuisine are *fan* and *cai*.[1] Like *yin* and *yang*, these are not opposites but complements. Balance is a key component of Chinese culture. Simply translated, *fan* means rice, and *cai* refers to dishes that accompany rice, including vegetables and meats. A balance of the two components is considered necessary to maintain health—except at restaurants or banquets, when the rice takes a backseat. Parents always ensure that their child eats a proper portion of *fan*. Rice is so vital to the concept of a meal in China that the term *fan* in a broader sense can be simply translated as "food," and the phrase *chi fan*, "to eat rice," also means simply "to eat."[2] Today rice provides a staple food source for nearly half the world's population.

Rice was unquestionably significant in the diet of ancient China, especially in the rice-growing regions of the south. To the north, where rice could not grow, winter wheat and the millets were the most important crops, and in the high elevations in the west, the main crop was barley. Much of the rest of the arable land in China was devoted to rice, or crop rotations that included rice. In the semitropical southern regions of China, two rice harvests are possible per year on the same tract of land.

Although rice has been a part of Chinese cuisine for millennia, it became prominent only during the last millennium. During the Song period, rice yields in most regions of China doubled or tripled, leading to a decline in the importance of millets.[3] This increase was achieved partly through more intensive farming campaigns and earthworks, but possibly more important was the introduction of a new variety of rice: the fast-growing and early-ripening Champa, which originated in the Champa region that is now part of Vietnam. Historians claim that the rice was brought to the Yangtze River region in China in 1011, by an envoy of Emperor Chang-tsung (998–1022).[4] The introduction of Champa rice made two rice-crop rotations in the southern regions almost universal. After one rice harvest, farmers would immediately sow another rice crop. (Many of these farmers theoretically also maintained smaller gardens for a variety of vegetables.) However, because of the poor cooking qualities of Champa rice, it did not become predominant in regions of southern China where concerns like drought or length of the growing season were not as pressing.

Crop-rotation cycles allowed for the great agricultural surplus that spurred empire formation in East Asia, yet we know little about when the practice was first implemented and what it looked like in prehistory. Fan Sheg-Chih's agricultural manual from the first century BC, which survives in fragments, referenced in later sources, explains many aspects of Han farming, including the widespread use of multicropping and crop cycles of winter wheat and millets; ridge cultivation in wet areas; rice paddies; and the treatment of seeds before sowing, using bone-based fertilizers.[5] Many of these practices later became necessary in the face of economic change and eventually were mandated by political reform.

More recent reforms in rice cultivation have been far more dramatic. Following the Green Revolution effort to develop hardier, more productive, and more nutritious crop varieties, in the 1960s and 1970s Chinese agricultural scientists embarked on a campaign to study hybrid rice. One of their greatest achievements was the development of a male sterile line of rice, preventing a plant from pollinating its own flowers and thereby forcing hybridization.

This invention is credited to Yuan Longping, who has become an almost mythical figure in the Chinese scientific community and a household name in China.

Close relatives of East Asian rice (*Oryza sativa*) were also domesticated around the globe. Another *Oryza* species, *O. glaberrima*, often referred to as African rice, was domesticated near the Niger River some two to three thousand years ago. East Asian wild rice (*Zizania latifolia*) is reported in historic accounts or ethnobotanical studies across East Asia, from Siberia to Malaysia, although there is some debate as to whether all these reports refer to the same species.[6] It can be either collected in the wild or cultivated in shallow water. The cultivation of the plant for its long grains in eastern-central China is well attested in ancient literary sources. Interestingly, by the Song Dynasty (960–1279), *Z. latifolia* was no longer cultivated for its grains and was instead grown as a vegetable for its stem. However, it is a rather unusual vegetable, as it becomes soft enough to consume only after it has been infected by a fungal smut.

North American wild rice (the perennial species *Zizania aquatic* and *Z. palustris*) has a long history as a food for Ojibwe groups in the northern United States and southern Canada. Although the grain has been intensively harvested in the wild for millennia, it has never been morphologically domesticated, because the harvesting methods, which involved beating the grains into canoes, did not select for brittle rachis, as sickle harvesting did for wheat and barley.

The Domestication of Rice

As with East Asian millets, the date of the earliest human cultivation of rice has been a subject of heated scholarly debate over the past decade. Several locations for *O. sativa* domestication have been proposed, including India, the Yangtze River area in China, south China, the southern slopes of the Himalayas, and the coastal wetlands of southeast Asia.[7] Genetic studies have been used to argue for both a single center of domestication and multiple centers.[8] Until recently the generally accepted theory was that there were two centers of domestication: one in eastern China along the Yangtze River, and the other in northern India. However, new genetic data suggest a different and interesting story. It is becoming increasingly clear that *Oryza rufipogon* was brought under cultivation in the middle and upper Yangtze River Valley of eastern China by the seventh millennium BC, but that a fully domesticated population of *O. sativa* was not well established and stable until roughly 4000 BC.[9] The earliest evidence for rice paddy fields in China also dates to around 4000 BC, and it is clear that paddies were well established by 2500 BC.[10]

As with wheat and other grains discussed elsewhere in this book, the morphological feature that indicates the first shift toward domestication in rice is the nonbrittle (or tough) rachis. This trait, which indicates the loss of the natural dispersal mechanism, is used to distinguish between domesticated and wild populations of grains in archaeological sites. It is generally accepted that a mutation of the *sh4* allele in rice plants cultivated by humans resulted in this initial trait of domestication, although other genes are associated with nonbrittle rachises in some rice populations.[11] Other traits associated with domesticated rice include increased grain size (although this is not uniform, and it is affected by several genes) and reduced lateral branching or tillering, resulting in an erect habit (possibly linked to the *PROG1* gene).[12]

Humans have clearly had a long and close relationship with the rice plant in China. Possible archaeobotanical evidence for people using wild *Oryza* in eastern China dates back over twenty thousand years and appears sporadically through the Pleistocene.[13] Hunter-gatherers in East Asia started harvesting wild populations of rice (*Oryza rufipogon*) on a more intensive scale in the lower Yangtze River Valley by at least eight thousand years ago.[14] The oldest solid evidence that people were regularly harvesting wild rice in

China comes from Shangshan (ca. 9000–5000 BC). Although only ten charred rice grains were recovered from the site, mostly from the later phases, impressions of the husks, or glumes, of rice grains were well represented in pottery sherds and in burned soil.[15]

As with the millets (and many other crops), microbotanical data have been used to argue for a much earlier domestication of rice than existing macrobotanical data support.[16] For example, phytolith data from Shangshan have been used to argue that rice domestication dates back to 8000 BC. This conclusion was disputed by the paleoethnobotanist Dorian Fuller and his colleagues, who argued, based on macrobotanical data, that the phytolith-based dates for rice domestication needed to be pushed forward by several millennia. This team of scholars spent years studying the origins of rice domestication and examined thousands of rice rachises to determine if they had smooth or rough breaks. They concluded that there was a shift towards nonshattering spikelets in rice after 5000 BC.[17]

In wild populations of rice, a certain percentage of the grains adhere strongly to their ears (i.e., they have nonbrittle rachises), but as humans asserted selective pressure on the plant populations through harvesting with sickles, they increased the percentage of the population with tough rachises. In 2009, Fuller and his colleagues published the first large-scale study of rachis bases with the intention of identifying early traits of domestication at an archaeological site. They illustrated a long-drawn-out pattern of domestication at Tianluoshan in Zhejiang Province in the Lower Yangtze region. This study reshaped the scientific community's understanding of rice domestication.[18] An archaeobotanical study of rice spikelet bases from the archaeological site of Kuahuqiao (6000–5400 BC), also in the Lower Yangtze region, showed that most of the rice found at the site was from a wild form.[19] In contrast, a similar study at Tianluoshan showed that the domesticated types in the earliest samples (ca. 4900 BC) were in the minority but had become considerably more prevalent over the preceding three centuries.[20] At the site of Hemudu, also in the Lower Yangtze area, plant remains are remarkably well preserved because of the site's waterlogged status. These remains show that people at the site were collecting a wide range of wild resources for food. One was rice, which appears to have been under cultivation by the beginning of the fifth millennium BC. At Tianluoshan, the economically significant plants that were recovered included rice, water chestnuts (*Trapa natans*), acorns (*Quercus*, *Lithocarpus*, and *Cyclobalanopsis*), bottle gourd (*Laginaria siceraria*), fox nut (*Euryale ferox*), jujube, and persimmon.[21]

Long-Grain Basmati and Short-Grain Rice

Anyone who has eaten rice with curry at an Indian restaurant and the sticky rice in sushi rolls knows that rice grains vary in shape. Although there are many varieties of domesticated rice, they fall mainly into two well-defined clades, or branches: *O. sativa* ssp. *indica* and *O. sativa* ssp. *japonica*—or Indian and Chinese rice. *Indica* rice is generally long-grained: the clade is exemplified by the well-known basmati rice. *Japonica* is usually short-grained and is sometimes referred to as pearl rice. Many locally grown Asian varieties are intermediate in size between *indica* and *japonica*. Many *japonica* grains become sticky, or glutinous, with cooking, although there are glutinous and nonglutinous forms of both *japonica* and *indica*. Some varieties of both clades have traits that make them suitable for growing in wet paddies, and other forms are adapted to grow on drained land.

Recent genetic research has shed new light on the origin of East Asian *Oryza* rice. While it has long been accepted that rice was domesticated in the lower Yangtze River Valley, there has been considerable debate over a possible secondary origin in northern India. The Indian rice lines either evolved separately from the Chinese lines or have been isolated from them for several millennia.[22] Genetic investigations of tough rachises, the key trait of domestication in both clades, seem to point to a single process of domestication;

however, the overall genetic differences between these two populations suggest that rice was domesticated independently in China and India. Because of these differences, the latest common ancestor of the two clades is dated to between eighty-six thousand and two hundred thousand years ago.[23] The genomes of the two clades have distinct aspects, demonstrating that they had different wild progenitor populations—*O. nivara* for *indica* and *O. rufipogon* for *japonica*.[24]

The story is further complicated by the fact that the two rice populations share several mutations that are key to domestication, which would suggest one domestication process, or monophyly.[25] In addition to possessing the *sh4* allele, which produces tough rachises, most domesticated forms of rice have the *rc* allele, which causes the familiar white pericarp (outer bran layers), as opposed to the red wild forms.

The discrepancies in the genetic data have been reconciled through the development of a novel model for the domestication of *O. sativa*. Geneticists explain the similar domestication mutations between the two populations through hybridization. This implies that *indica* and *japonica* crossed at some point after the morphological domestication of *japonica*, leading to an introduction of the key domestication genes from the Chinese line into the Indian line.[26] According to the currently accepted narrative for *Oryza* rice, morphologically wild *O. nivara* was already being maintained and possibly cultivated by low-investment farmers or hunter-gatherers in northern India when domesticated *O. sativa* spread into the region from East Asia.[27] The cross between the two lines transferred the genes for tough rachises and other domestication traits into *O. nivara*, effectively domesticating it and leading to its being grouped taxonomically with *O. sativa*. Several key genes crossed the species barrier, but the most important was the *sh4* allele.[28]

Rice was an established part of the economy in the Ganges region of India by the middle of the third millennium BC, and archaeobotanical finds of the grains have been reported at domestic sites in Punjab, Harayana, and

Swat.[29] Scholars have argued that domesticated rice was present throughout the same region by the very beginning of the third millennium BC.[30] The crop was well established in much of India and southeast Asia by about 1500 BC.[31]

Some of the earliest domesticated-form rachises for *indica* rice, dating to the beginning of the second millennium BC, have been recovered from the archaeological site of Mahagara, India.[32] This discovery implies that the two cultivated lines hybridized before that time. This hypothesis fits with the proposed spread of cultivated wild *indica* rice from the Ganges River region to the Upper Indus River Valley by the third millennium BC. Archaeobotanical data from Lahuradewa on the Upper Gangetic Plain of Uttar Pradesh, India, has shown that wild rice was being harvested and possibly maintained in the region as far back as 6000 BC.[33]

It is still not clear exactly by what routes domesticated *japonica* traveled to meet up with wild *indica*. A route across southeast Asia or possibly along the southern Himalaya through the Swat Valley to the Indus seems like the most logical answer, but an alternative hypothesis has recently been presented. A team of paleoethnobotanists from University College London suggest that the crop could have moved along the northern routes of the proto-Silk Road, accompanied by many of the other crops discussed throughout this book. While there is no solid archaeobotanical evidence for rice in northern Central Asia before the common era, it is still a fascinating hypothesis. The authors suggest the possibility that rice followed the same routes as millet on its path from East Asia through Central Asia. They acknowledge that further investigation is needed in northern South Asia and throughout the Himalayas, especially in light of several new finds of early rice remains in the southern Himalaya.[34]

Archaeobotanical data do not show conclusive evidence for riziculture west of the Central Asian mountains until the first century AD. Pilov, or pilaf, is possibly the most widely shared dish of the Russian, Turkic, and

Arabic worlds today, but it likely originated in some form over two millennia ago in India and gradually spread across Russia. This dish takes many forms: it contains rice, dried fruits or carrots, onions, and sometimes meat, and every region and household tweaks the recipe in its own way. Pilaf has become a symbol of identity and on occasion a powerful statement of nationalism. Many Central Asians take great pride in their local versions of the dish. Interestingly, although it is almost impossible for modern diners to imagine this dish without rice, until relatively recently only the richest members of society could afford to make it with rice: everyone else used barley.[35]

The Westward Journey of Rice: Reshaping Turkic Cuisine

Like the millets, Asian rice first evolved in eastern China but eventually became prized in cuisines far to the west. Rice is indispensable in Arabic and Turkic cuisine today, and it was a significant part of the diet at least as far back as the medieval period.[36] Persian, Arabic, and Islamic cuisines cook rice in oil or steam it and serve it with a wide variety of vegetables, spices, and meats. Rice also featured in the diet in other ways: it was an important component of medieval Arabic desserts, rice flour was used to make breads, rice was fermented into beer and vinegar, and it was used medicinally. However, among most Central Asian cuisines today, its starring role is in pilaf.

According to Berthold Laufer, the leading authority in the early 1900s on plant exchange along the Silk Road, rice was not cultivated in West Asia before the fourth century BC.[37] Literary sources and archaeobotanical evidence show that rice was known in southwest Asia as early as the fifth century BC.[38] However, it does not appear to have become a significant crop until after the Islamic conquests and the establishment of more elaborate irrigation systems. It is mentioned in a few Classical texts, often as an oddity, and was likely not grown in any abundance in the Mediterranean basin until Arab merchants became more established in eastern regions after the seventh-century

Islamic expansions. The physicians Dioscorides and Galen (AD 130–210) are supposed to have referred to medicinal uses of rice. The cookbook *Apicius* (fifth century AD) and the Byzantine physician Anthimus (sixth century) refer to rice as growing in lands to the east. *Apicius* mentions rice cooking water as an ingredient used in other dishes or medicinally, suggesting to some historians that rice had culinary as well as medicinal uses in ancient Rome.[39] Rice does not appear in the Avesta, the ancient Zoroastrian sacred texts (which mention only a handful of plants); nor is it among the grains that Herodotus mentions as growing in Persia.

The only early, well-dated and well-documented find of archaeological rice grains outside East Asia and India comes from work by the archaeobotanist Naomi Miller. She recovered 373 grains of a short-grained variety of rice from a cache or offering on the floor of Level 3A at Ville Royale II, at Susa, in Iran, in association with a jar that was likely used to store the grains.[40] Her findings are consistent with an observation by Strabo that rice was grown in "Babylonia," Bactria, Susa, and lower Syria.[41] Strabo did not travel to these regions himself; he is likely referencing the accounts of Aristobulus, who traveled with the army of Alexander the Great and hence would have made the observations sometime between 334 and 323 BC. Aristobulus presumably followed Alexander all the way to the Indus: he mentions rice paddy farming, which he may have seen in Punjab.[42]

Support for Strabo's claim comes from Diodorus Siculus's *Bibliotheca Historica*, which notes that rice was grown in India as a summer crop along with millet and sesame, whereas wheat was grown in the winter.[43] Diodorus Siculus may be reiterating comments from earlier texts and probably did not travel to India personally, but he does appear to recognize rice.

Assuming that Strabo and Aristobulus were indeed referring to Asian rice, we can assume that by Hellenistic times, rice was known in the Mediterranean. This view is supported by an observation by Theophrastus that rice was grown in India, which indicates that he at least knew what rice was.[44] Dioscorides mentions making bread from rice flour, and Pliny the Elder refers to rice several times in the *Natural Histories*.[45] Archaeobotanical remains of thirty-three grains of rice from the second century AD were recovered from the Roman trading port of Quseir al-Qadim in Egypt, and a few additional grains were recovered from excavations at the Roman trading center of Berenike. The inference from these very small finds was that the grains were imported; interestingly, at both sites rice appears to have remained in low abundance throughout the Islamic period.[46]

According to the *Shiji* (The records of the Great Historian), the Han traveler Zhang Qian reported that rice was cultivated in the land of Dayuan.[47] Most scholars agree that this refers to Fergana in modern-day Uzbekistan. As part of his report on the local culture, the military ambassador recorded the main grains of the new region, taking special care to observe whether rice was cultivated, and noting in addition that wheat fields and grape vineyards were abundant in Fergana. Zhang Qian also claims that rice was being grown in Parthia and Chaldea, although many scholars have pointed out that these are not firsthand observations.[48]

Archaeobotanical studies in Central Asia have not recovered solid evidence for rice, but these data are very limited, especially for agriculturally rich regions like Fergana. A 1970s Soviet excavation claimed to have found a large quantity of rice grains at one site in the numbers 28, 29, and 61 group in Fergana, dating to the early first millennium AD. Although this claim has not been properly verified, it may be supported by a second report from the 1980s of rice grains recovered from mudbrick fragments at Munchak Tepe, in the Osh region of Kyrgyzstan near the town of Kerkidon, dating from the fifth to the seventh century AD.[49] Rice is grown today in parts of the Zerafshan valley in Uzbekistan and in some humid river valleys as far north as southern Semirech'ye, but it is unlikely that rice was grown in most of this area in the past.

A more enticing report comes from a small archaeobotanical study conducted at the medieval villages of Djuvan-tobe and Karaspan-tobe, both on a tributary of the Syr Darya in southern Kazakhstan. The botanist working on the project identified a single rice grain from fourth- to fifth-century AD occupation layers at Karaspan-tobe and another single grain from the seventh century at Djuvan-tobe. These discoveries of only two grains stand in contrast to finds of several hundred grains of wheat, barley, and broomcorn millet from the same sites. The scholar does not provide descriptions or images of the two grains, despite providing wonderful illustrations of all the other domesticated crops found at the site, including apple seeds, grape seeds, peas, and lentils.[50] He supports his claim by mentioning that excavators at a site in the same region of Kazakhstan, the village of Konyr-tobe in the Otrar oasis, dating to the seventh century AD, also discovered a small quantity of rice.

Taken together, all these reports of rice, while individually questionable, suggest that rice was becoming more prevalent in Central Asia by the middle of the first millennium AD. However, it was probably still a minor grain, with wheat, barley, and the millets taking a far more prominent role in daily cuisine.

Phytolith analysis at the ancient village site of Tuzusai in Semirech'ye yielded results suggesting the presence of rice. Arlene Rosen and her colleagues reported finding silica skeletons of cultivated rice husks (the shell that covers the grain) as well as "fan-shaped keystone" phytolith forms that they claim are indicative of the leaves and stems of rice plants.[51] However, after many years of macrobotanical research at the site and additional phytolith work, all attempts that my colleagues and I have made to confirm these findings have been unsuccessful.[52] A large number of species of grass produce a similar fan-shaped bulliform phytolith, notably *Miscanthus* and *Phragmites* spp., and both genera grow near the site today. Furthermore, fan-shaped phytoliths have been found in other areas of Central Asia where rice was far less likely to be cultivated, notably in desiccated sheep or goat dung from a 400 BC tomb in Xinjiang.[53]

Given that the only reports of rice in the Central Asia area before the Islamic expansion come from highly problematic analyses of phytoliths and a few unverified Soviet-period archaeological reports, it seems hard to support claims of earlier rice cultivation. While it is likely that rice was moving into southwest Asia by the latter part of the first millennium AD, it probably did not become established in other parts of Central Asia until the second millennium AD. The grain that plays such a vital role in Turkic and Arabic cuisine today apparently played a minimal role on the early routes of the Silk Road.

One of the most interesting macrobotanical finds of rice in the past few years comes from Kyung-lung Mesa (or Kaerdong) in Tibet. The recent discovery came from layers dating from between AD 455 and 700. It consists of only one rice grain but also nine rice spikelet bases, and it is supported by another find of rice at a roughly contemporaneous nearby site called Zebang. Spikelet bases usually imply local cultivation, but the Kyung-lung site is located at an altitude of 4,300 meters and Zebang at 3,000 meters, well above the normal growing zone for rice. Whether the grains were grown locally or carried more than a thousand kilometers from the valleys to the south, the presence of rice at these sites suggests that it may have spread into the southern Himalaya well before it was established in southwest Asia. It is also possible that rice followed a route through mountain passes in the southern Himalaya to reach Central Asia.[54] This theory is supported by finds of rice at Semthan in Kashmir dating from between 1500 and 500 BC.[55] Rice may not be the only East Asian crop to have passed through Kashmir (notably the Karl Gandaki valley): scholars have pointed out early evidence of peaches, apricots, and Job's tears (*Coix lacryma-jobi*).[56] Rice grains have also been recovered from burial offerings from the Astana cemetery in the Turfan of Xinjiang, dating to the Tang Dynasty.[57] The Astana rice remains are the earliest clear evidence of rice in northern Central Asia and along the main arteries of the Eurasian exchange routes.

In the early 1900s, Berthold Laufer argued that rice did not become prevalent in southwest Asia until the beginning of the Arabic period, after the seventh century AD.[58] More recently Andrew Watson noted, based on historical sources, that "in the eastern part of the caliphate, where rice had been grown in ancient times, the early centuries of Islam saw an extension of its cultivation." He further argued that with the development of irrigation systems after the Muslim conquests in the early seventh century, "rice early came to be grown in the Islamic world almost wherever there was water enough to irrigate it."[59] The thirteenth-century cookbook *Scents and Flavors the Banqueter Favors* specifically mentions rice in connection with nine dishes, mostly pilafs, but the book was written for the elites, and rice is a minor component in it. By comparison, it presents dozens of ways to make pickled vegetables.[60]

Watson mentions early Islamic sources attesting to the cultivation of rice as far west as modern Russia, south across northeastern Africa, and north as far as Fergana. We know from historical and archaeobotanical sources that it was cultivated across all of East Asia by this period. He also references several early Islamic texts that discuss paddy-style agriculture, clearly noting that dry rice cultivation was not implemented in West Asia.[61] Nonetheless, well into the Islamic period, rice remains uncommon among archaeobotanical remains. In studies of material from Islamic village sites in the Upper and Middle Euphrates region, no rice was recovered from the eighth- through tenth-century sites, but small amounts of carbonized grains and rachis bases were recovered from eleventh- through fourteenth-century contexts.[62] After AD 1000, rice was cultivated throughout inner Asia, even in areas that now seem too dry to support it, like most of Afghanistan.

Rice is better suited for cultivation in the humid, semitropical regions of South Asia than the desert oases stretching from modern-day Afghanistan to Syria, where its cultivation incurs high labor and ecological costs. In addition to requiring extensive labor for irrigation, rice is likely to increase salinity in soil, especially if irrigation water is left in a field to stagnate. Water seeping through the soil dissolves mineral deposits below the surface, allowing them to slowly percolate to the surface and become concentrated as the surface water evaporates.

Many medieval Arab sources discuss practices for rice cultivation, especially ways to work it into crop-rotation cycles; one of the more detailed of these discussions was provided by Ibn al-Awam. Another detailed account is provided in the *Book of Nabatean Agriculture*.[63] One method for reducing salinity was, and is, to drain and replenish the standing water, but this further increases the ecological demands of this crop. Ethnohistorical sources from northern Iran note that rice fields were kept small and that water had to be circulated continually between field catchments to avoid high saline levels. In addition, fields were often left fallow for three years between rice harvests and used for grazing instead.[64]

The archaeobotanist and current curator of the ethnobotanical collections at Kew Gardens, Mark Nesbitt, along with two of his colleagues, compiled an exhaustive summary of the textual and archaeobotanical evidence for the spread of rice into Central Asia. They note that the Babylonian Talmud (third to fifth centuries AD) mentions rice consumption near the Persian Gulf and the alluvial plains of Mesopotamia, and that the crop was locally significant during the Late Sasanian period and was subject to taxation. Rice was included in descriptions of royal dinners during the Sasanian period, notably served as a jelly. However, it apparently remained a rarity, reserved for the elites and grown only in small areas where conditions permitted its cultivation without significant labor investment in irrigation systems.

The only other archaeobotanical remains of rice from Western Asia date from between the ninth and the twelfth century, including a few grains recovered from one site in Syria and one in Turkey (both on the Euphrates).[65] From the Abbasid period onward, rice cultivation gradually increased as irrigation systems were improved and culinary tastes changed. Historical sources emphasize the use of rice flour for bread baking, although the flour seems to have been used to make sweet pudding or jelly dishes as well.

This transformation in the culinary role of rice as it moved westward is reflected in the spread of other crops. For example, millet, which was consumed as gruel in China, was baked into hard bread in Central Asia; likewise, wheat, which was used for making bread in Central Asia, was turned into noodles and steamed buns in China. In Turkic and Arabic cuisines, however, the large-scale adoption of pilaf may reflect very recent cultural changes.

One of the tenth-century Arab geographers, al-Muqaddasi (AD 945–91), refers to major expansions in government-sponsored irrigation projects in the Kur River area of the Fars region of Iran and notes that the ancient city of Istakhr on the Pulvar River, a tributary of the Kur, was surrounded by rice fields and fruit orchards.[66] Other Arabic scholars also mention the growing importance of rice. Rashid-al-Din Hamadani (1247–1318) was a Sunni statesman, a prolific writer, a historian, and a physician who worked with a team of great scholars, including Buddhists in Iran during the Ilkhanate period, and became the vizier of the Ilkhan ruler Ghazan. He describes a variety of rice that was popular in India, which historians identify as basmati, but he also notes that the Persians failed to grow it in southwest Asia because of ecological constraints. Some historians have suggested that this was the period when rice really became popular in southwest Asia, especially with the large dam-building projects undertaken in the Mongol period.[67]

Around the same time, Sayyar al-Warraq compiled one of the earliest of the Arabic cookbooks, containing 615 recipes that he pulled from other contemporary sources. Only a few of these recipes involve rice, often in the form of bread or beer; there are also a few recipes for rice porridge spiced with cassia, sugar, ginger, or honey. Three other important Arabic cookbooks date to the thirteenth century; the one written slightly later has more recipes for rice.[68]

A more detailed study of Arab sources that discuss the cultivation and uses of rice is presented by Delwen Samuel, an archaeobotanist at the

Institute of Archaeology and University College London. He notes that historical sources from the ninth and tenth centuries talk about the use of rice not only in savory dishes such as soups but also in sweet puddings, cooked with milk, fermented, or baked as a bread. These dishes are sometimes sweetened with fruit syrups, saffron, sugar, grapes, figs, dates, or honey.[69]

Closer to the modern period, texts such as the *Memoirs of Babur* attest to the importance of rice in Central Asia more than five hundred years ago. Babur described large rice fields in the valleys of the Kafiristan region of Afghanistan. As his troops moved through the region, causing most of the Kafir people to flee and killing those who remained, his armies seized large quantities of rice.[70] The sixteenth-century account of the life of the Emperor Akbar, written by his vizier, Abu'l-Fazl ibn Mubarak, provides numerous recipes for dishes prepared with rice.[71] These texts suggest that rice was well known across the Islamic world by the eighth century; however, it was probably not a major part of the cuisine, especially among commoners, until the thirteenth century.

Oryza sativa: Conclusions

The currently accepted model for the domestication of rice suggests that people in eastern China started collecting wild *Oryza* rice as early as 6000 BC and may have started applying genetic pressure on the wild plant population. Over the following millennia, anthropogenic pressure led to significant changes, notably selecting for nonbrittle rachises through the harvesting process. Eventually, foragers in the Lower Yangtze River Valley, who were collecting a wide range of fruits, nuts, and seeds, inadvertently genetically altered the population of wild rice around them. The results of this process are visible in archaeobotanical remains at the sites of Tianluoshan and Hemudu. The gradual transition from wild to domesticated rice at Tianluoshan did not culminate until around 4600 BC.[72]

Following a parallel track, rice populations in parts of northern India were being harvested from the wild by foragers as far back as 6000 BC. The selective pressures on Indian wild rice did not lead to morphological changes. Nonetheless, fully domesticated rice, genetically distinct from the rice in East Asia, was being cultivated by the early second millennium BC at Mahagara. The genetic data suggest that rice was independently brought under cultivation in both the Yangtze region and in northern India, but that the allele for nonbrittle rachises, the trait of domestication, originated in the Yangtze region and later spread to northern India.[73] Therefore, the morphological domestication of East Asian rice occurred only once, in the Lower Yangtze River Valley, and the genes for domestication were transferred from fully domesticated *japonica* rice to previously wild *indica* rice—resulting in what we know today as basmati and pearl rice. The gene later spread with the crop and became dominant (introgressed) in other populations of rice that were previously morphologically wild but cultivated in northern India.

It has been proposed, albeit on the basis of piecemeal data, that the genes for domestication spread to the wild *indica* populations along northern routes of interaction, through the mountain foothills of Central Asia.[74] New data seem to suggest that it spread through Kashmir, possibly down the Swat Valley to Punjab. Interestingly, even though rice is an essential part of cuisines in Central Asia today, it probably did not become a significant crop west of the Indus Valley until the Islamic conquests of the seventh century AD. Therefore pilaf, the rice-based dish that now exemplifies many Central and southwest Asian cuisines, originated in that part of the world only in the past millennium. While rice is today one of the world's most important cereal crops, the history of its spread across Eurasia is shorter than most people might think.

BUCKWHEAT

Buckwheat is not related to wheat; it is not even a grass. It is a member of the knotweed family, Polygonaceae, which has attracted the attention of humans around the world for millennia. The common English name is an

Anglicized form of the German *Buche-weizen* (beech wheat or grain), referring to its similarity to the nuts of the beech tree (*Fagus* spp.): thus the Latin name also alludes to this resemblance. At least one close relative was cultivated in North America—erect knotweed (*Polygonum erectum*). Species from the genus *Fagopyrum* were domesticated by early Asian populations at least three separate times. The best known of these domesticated buckwheat crops is common buckwheat (*Fagopyrum esculentum*), which is believed to have been domesticated in far western China. A second domesticated species is Tartary buckwheat (*F. tataricum*), also known as bitter buckwheat or duckwheat. *F. tataricum* can grow at higher elevations than *F. esculentum*, up to 4,300 meters. Unlike other *Fagopyrum* species, *F. tataricum* is frost tolerant. Many scholars have suggested that this crop originated in the mountainous regions of western China.[75] Although far less common, *Fagopyrum acutatum* (Japanese or silverhull buckwheat) was once grown in Japan and far East Asia.

Buckwheat is a staple high-elevation crop of the Himalaya, and it costars with barley in Tibetan cuisine and farming. It is tolerant of short growing seasons, cold climates, poor rocky soils, and limited maintenance.

Today, buckwheat appears in American and Western European diets only in pancakes and a few other specialty baked goods. In Eastern Europe and Russia, it is a common porridge grain. G*rechka*, the Russian buckwheat porridge, has a strong flavor and is commonly served to children with sugar. Yet the limited archaeobotanical data seem to suggest that the grain probably did not spread out of the Himalaya or East Asia until well into the common era. Even then, *Fagopyrum* never became established in most of Central Asia or southwest Asia, and its use as porridge in northern Central Asia may be a consequence of Russian conquest.

Relatively little is known about the origins and spread of the Polygonaceae. Because of its importance in the Himalayan regions, as well as the huge diversity among landrace varieties and wild relatives of the crop in this region, scholars assume it was domesticated somewhere in that area. There are, however, no archaeobotanical assemblages that testify to predomestication cultivation or the domestication of buckwheat. Genetic research on wild stands of *Fagopyrum* in Tibet suggests that the wild progenitors of common buckwheat (*F. esculentum* ssp. *ancestrale*) and Tartary buckwheat (*F. tataricum* ssp. *potanini*) may both have originated on the Himalayan Plateau. This would imply that common buckwheat was domesticated in eastern Tibet or the Deqin district of Yunnan and that Tartary buckwheat originated in central Tibet.[76]

Buckwheat was brought to Central Asia only within the last couple of centuries, likely as a result of Russian imperial expansion. It is notable by its absence from the descriptions of farming practices discussed by early-nineteenth-century European explorers of the region, though many of these, like Eugene Schuyler, go into great detail about the crops they saw under cultivation.[77] When Chang Chun passed through Samarkand in 1222, he noted that all grains and legumes that grow in China also grow in the fields around the city, with the specific exceptions of buckwheat and soybean (assuming that Emil Bretschneider's 1888 translation is correct).[78] The only mention of buckwheat by a European explorer in the region comes from Henry Landsell, who was fascinated by the high-elevation farming practices he observed in the Ili River Valley and on the slopes of the Tien Shan Mountains. He mentions the cultivation of buckwheat, along with other grains, in the environs of the town he refers to as Vierny.[79] This location is difficult to identify with certainty. It may be Almaty, but Schuyler visited a town he called Vierny in the Ili Valley about a decade earlier and noted its elevation of 2,400 meters, whereas modern Almaty is only at 800 meters.[80]

While there is no solid evidence for buckwheat's having spread out of East Asia more than a thousand years ago, there are a few contentious studies, based on microbotanical data, that have argued for its spread into Europe as early as the fifth millennium BC. For example, a study using pollen data traces the spread of buckwheat in great detail from the southern Himalaya north along the mountain corridor and ultimately across the northern Central Asian steppe around the fifth millennium BC.[81] However, archaeologists and macrobotanical specialists have pointed out that there are almost no other data and no good macrobotanical evidence to support these conclusions.[82] A few other claims of buckwheat's early arrival in Europe can mostly be dismissed. Other studies using pollen and starch grain data from China to argue for early buckwheat dates are clearly problematic (see below).[83]

Scholars as far back as the botanist Alphonse de Candolle in 1884 have pointed out that there are no early literary references to the grain in Europe and that the linguistic evidence suggests a late arrival. There are no early references to the crop in southwest Asia, and there is no Sanskrit word for buckwheat; nor does it ever appear in ancient Greek or Roman literature, and the taxonomic name is a modern invention, arising from the lack of a Latin name for the crop. Furthermore, there are no common root names for buckwheat in any European languages. De Candolle suggested that the crop spread to Russia through far northern Central Asia (the region he calls Tatary) in the medieval period and was not recorded in Western Europe until 1436, when it was planted in Germany.[84]

It is fascinating that a crop that became significant in Europe appears to have such a recent history and mysterious route of spread. Its history in East Asia is similarly enigmatic. A published account of one grain was reported from early Jomon layers in Japan, thought to be roughly five millennia old, but this is generally accepted to be an intrusion from later layers.[85] Buckwheat is not mentioned in early Chinese historical sources until the fifth and six centuries AD, and there have been almost no archaeobotanical finds of the grain in central China.[86] A recent synthesis of the macro- and microbotanical data for buckwheat in China illustrates just how mysterious its origins are. The authors compiled twenty-six claims for the early appearance of buckwheat across China, fourteen of which were based on pollen data and two on starch grains. The remaining ten claims for early seed remains require further exploration. As with much of the microbotanical data for early agriculture in East Asia, the dates from the starch grains significantly precede any reliable dates from macrobotanical data and are highly spurious. The buckwheat pollen data from China are just as problematic as the pollen data from Europe, and the identifications of domesticated forms of *Fagopyrum* have been called into question by one deep sediment core, taken using a drilling machine, from China, with dates for the plant reaching back to 23,000 BC. Likewise, *Fagopyrum* is an indicator of disturbed ecologies, and it is a plant that spreads its seeds by means of grazing animals eating the whole plant, seeds and all; hence, changes in the frequencies of pollen from this genus in the pollen record are more likely to reflect herd grazing practices than human activity. The oldest reliable sample is from the middle of the first millennium BC in Yunnan. Slightly later macrobotanical remains have also been recovered from Qinghai, Shaanxi, and Gansu.[87]

The oldest likely finds of cultivated buckwheat come from Yunnan, China, where three small-seeded specimens were preserved from the site of Haimenkou. The seeds, which date to roughly 1400–800 BC, may or may not come from wild buckwheat.[88] There are other claims of buckwheat from southern China dating to 1500 BC, but follow-up analyses are needed.[89]

More clearly domesticated remains of buckwheat grains were found at cemetery sites at Mebrak and Phudzeling in the Jhong valley in Upper Mustang, Nepal. The grains were recovered from burials dating from between 1000 BC and AD 100, clearly illustrating that the crop was cultivated in the high mountains by the end of the first millennium BC at the latest.[90] However, debate over buckwheat's origins and spread continues, and at least one team of archaeobotanists is still arguing for a center of origin in northern China around 2500 BC.[91] Until solid supporting evidence is produced, these early dates must be viewed as speculative; nonetheless, we should not rule out the possibility that *F. esculentum* originated somewhere farther north and outside of the mountains. Interestingly, the archaeobotanical report from the Nepalese sites notes the presence of both buckwheat species (*Fagopyrum tataricum* and what the authors think is *Fagopyrum esculentum*). By this time, the southwest Asian crop assemblage had already spread into the region, and people at Mebrak and Phudzeling were also growing naked and hulled barley and flax.[92] Most likely, buckwheat originally grew as a weed in these early wheat and barley fields but was gradually domesticated through human tolerance of its presence. It is also possible that dense stands of this plant existed as a result of heavy yak herding; the presence of ruminants often leads to concentrations of endozoochoric plants (those dispersed by animals that consume them). Ultimately, buckwheat domestication seems to be a secondary result of well-established agropastoralism in the southern Himalaya.

One of the most promising keys to understanding the domestication of buckwheat comes from macrobotanical finds from the site of Kyung-lung Mesa, from occupation layers dating between 455 and 700 AD.[93] The archaeobotanists who discovered the two possible specimens of Tatary buckwheat concluded from their morphology that they were domesticated. Earlier work at the same site conducted by me and several colleagues found *Fagopyrum* sp. seeds, but we were not confident enough on the basis of the few specimens that we recovered to call them domesticated.[94] Preserved buckwheat grains dating to the first millennium AD were also recovered from the site of Kohla in the mountains of Nepal.[95]

On the basis of current archaeobotanical data, it seems most plausible that the crop was a late domesticate, brought under cultivation in the southern Himalaya as recently as the second millennium BC, after other crops were already being cultivated in the region. It was likely a weed before it was cultivated. The scant available data suggest that the two mountain species were not actually domesticated until about three millennia ago. Likewise, buckwheat likely did not spread along the Silk Road to the north or into the heart of China until the first millennium AD, and into Europe and Central Asia later than that. The spread of the crop into Russia proper and thence to Eastern Europe was arguably a result of the formation of a tea trade route into Moscow, which was the political center of the Russian world by the thirteenth century (see chapter 12).

THE LOST GRAINS OF ANTIQUITY

For at least ten thousand years, ancient Asians have fostered a close coevolutionary relationship with grasses and pressured a surprisingly large number of species into evolving domestication traits. Many of these domesticated Asian grains were unknown elsewhere, and some have faded into the annals of history. In addition to the familiar large-seeded grains, such as wheat, barley, rye, oats, rice, and broomcorn millet, Asian domesticated grains include pseudocereals in the Amaranthaceae family, such as *Chenopodium* spp. (mentioned in chapter 11); the Polygonaceae family, such as the buckwheats (*Fagopyrum* spp.); and over a dozen species of grass (Poaceae). Few of these lesser grains ever made their way onto the Silk Road.

Many other small-seeded millets originated in East Asia besides broomcorn and foxtail millet. The genus *Echinochloa* was domesticated repeatedly in East Asia, most notably in Japan, where its domestication dates back at least four millennia. Japanese millet (*E. esculenta*) was cultivated on a limited scale in Japan, Korea, and parts of eastern China, often in dry areas that are too cold for rice cultivation. Another *Echinochloa*, *E. frumentacea* or Indian barnyard millet, was domesticated in South Asia and grown in India, Pakistan, and Nepal. A third domesticated species, burgu millet (*E. stagnina*), hails from West Africa. Wild or feral *Echinochloa* species are still grown or allowed to grow in millet fields in Uzbekistan today. The grains from these crops are harvested along with the cultivated millets.

A number of other small cereals were domesticated in South Asia, such as *Brachiaria ramose*, *Panicum sumatrense* (little millet), and *Setaria glauca*, all originating somewhere in southern India. Historically, *Panicum sumatrense* and *Setaria glauca* were grown together in mixed stands in southern India.[96] Another interesting East Asian cereal, which was widely cultivated in parts of China in the past, is Job's tears, which still grows wild or feral across central China. This is a unique grain with a hard shell. Examples have been recovered from the Xinjiang burial ground of Sampula, dating to the early first millennium AD.[97] The grain has been collected from the wild at various times in history.

Some other southwest Asian grains made their way into China but had a less profound influence on cuisine. These include rye, called *hei mai* or black *mai* in Chinese (*mai* being a common general term for wheat, barley, and other grains originating in the west), and oats (*qiao mai*, or sparrow *mai*). Rye was traditionally cultivated in Sichuan and Yunnan, but almost nothing is known about its introduction or its spread into East Asia.[98] It is better attested in southwest Asia, where it has been identified at the twelfth- through fourteenth-century site of Qaryat Medad, and one rye grain dating to the same period was found at the site of Tell Guftan, Syria. Both assemblages contained large quantities of wheat and barley grains.[99] Almost nothing is known of the spread and early cultivation of oats in China, although the grain has been grown in western Sichuan by the Lolo people and in southwestern Sichuan by the Nosu (black Lolo) people.[100]

Another grain that came to play an important role in China is sorghum, which was domesticated in northern Africa and likely entered China from a southerly route. The old word for sorghum in Chinese was *shu-shu*, or Sichuan millet, and some linguists have argued that this term may indicate a link to an overland route from India.[101] In the absence of any evidence for the crop in Central Asia or in historic Turkic cuisine, this hypothesis seems plausible. It is also unclear when sorghum spread into southwest Asia, but Pliny stated that a "kind of millet has been introduced from India into Italy within the last ten years, of a swarthy colour, large grain, and a stalk like that of the reed. This stalk springs up to the height of seven feet, and has tufts of a remarkable size, known by the name of 'phobae.'" It seems likely that he was referring to sorghum.[102] Historical sources in China do not mention *shu-shu* until AD 300, and there is disagreement over whether it might have arrived earlier.[103] Like the other millets, sorghum is hardy and tolerant of summer heat and aridity. It is widely grown in southern China today and a common source of sugars for producing *baijiu* liquors. But the lack of solid early archaeobotanical evidence for the crop in China likely suggests that it did not move north from India until the common era.

We know little about landrace varieties of crops in the past or how much diversity existed in farmers' fields. In one unique case, entire ears of broomcorn millet and cereals were found preserved in the 1933 excavations at the fortification of Mugh, Tajikistan, giving us insight into the varieties of crops that were grown in Fergana and the Zerafshan regions during the early medieval period. The barley was a hulled form with long grains that had long awns, or barbed hairs, extending from their tips, similar to varieties used for animal fodder across Central Asia today. The wheat from the Mugh fortified citadel was awned and appears to have been a variety of wheat more commonly used for baking breads than for making noodles. In addition, the original report claims that it was a red-wheat form, with red awns, glumes (hard shells that enclose and protect the grains), and grains. Historically, red wheats were cultivated across southern Central Asia and parts of China. In addition to ears of broomcorn millet, ears of *Echinochloa crus galli* were found. This wild millet relative is a problematic weed in fields across Asia. In the Zerafshan region today, I have seen it grown intermixed with broomcorn millet and harvested along with it. However, some scholars point out that it was once cultivated in Uzbekistan and Tajikistan under the local name of *kurmak*.[104]

CHAPTER SIX

Barley

HULLED OR NAKED VARIETIES: BEER OR BREAD?

Barley, like broomcorn millet, has been consumed by peasants across Eurasia for millennia, although its culinary and social roles have changed greatly over time. In ancient Greece and Rome, the cereal had religious significance (notably as a symbol for Demeter or Ceres, the goddess of agriculture). In the *Iliad*, sacred barley meal is sprinkled over recently bled animal offerings, and in the *Odyssey*, white barley flour is served as an offering to appease the dead. Barley bread was used in ancient divination rituals such as alphitomancy (a means of identifying the perpetrator of a crime) and in a similar way in the Anglo-Saxon practice of corsned, or morsel of execration.

Because of its hardiness, barley allowed farmers to settle some of the world's highest and least arable mountain ranges and produce yields in soils where wheat would fail. Today, barley's most important role worldwide is in the production of alcoholic drinks. Traditionally, beer in Europe and America has been produced from barley malt flavored with hops (*Humulus lupulus*). It is also distilled into bourbon, Scotch, and Irish whiskey. Barley is still used in the Euro-American culinary tradition to thicken soups (as in beef barley stew) and occasionally to produce heavy breads; however, with the industrialization of agriculture and the mass production of bread, wheat flour has largely replaced barley meal.

Barley has been an indicator of social standing in many societies. In the *Epic of Gilgamesh*, Gilgamesh eats barley cakes in an effort to associate more closely with the peasantry. Plato notes in *The Republic* that in the ideal *polis*, citizens would be fed on the dense loaves of the rural class rather than the lighter loaves of city dwellers. A similar polarization is visible in the writings of the Greek physician Galen (AD 129–ca. 200), who noted that Greek armies were fed barley porridge, whereas Roman armies considered eating barley to be a punishment.[1] Bread wheat, especially the higher-gluten varieties developed for bread baking, produces a lighter and more palatable bread; however, as a crop, it requires more water and is less hardy than barley. In addition, barley requires less labor and is better adapted to cold climates and northern latitudes. These traits made the crop the staple of European pioneers and settlers during the sixth and fifth millennia BC, although it was not until the development of frost-tolerant lines and a loss of photoperiod sensitivity (discussed below) that barley truly took hold in Eurasia.[2]

As with wheat, there are two main categories of domesticated barley: naked and hulled. The individual grains of hulled varieties of barley have a hard protective shell or glume. These glumes need to be removed through processing, usually beating and winnowing. The naked forms have a genetic change that makes their glumes thin and easily removable. Naked barley requires more water but is more easily processed into food. Hulled barley requires less labor during cultivation (and often no irrigation) but requires more labor in order to be ground into a flour or cooked and eaten as groats. Therefore, hulled forms of barley are often used for purposes such as fermentation or animal feed, for which milling is unnecessary.

These differences help map changing culinary and agricultural practices during the early period of European farming. Two archaeobotanical experts from Cambridge University, Diane Lister and Martin Jones, examined a large data set of archaeobotanical reports from across Europe. They concluded that during the fifth and fourth millennia BC across the Caucasus and the Mediterranean, the cultivation of hulled grains gradually gave way to naked varieties, although assemblages containing both were still prevalent in Eastern Europe.[3]

The preference for naked grains may have arisen out of a preference for growing the more easily cultivated barley rather than wheat. In many parts of Europe, a respectable crop of barley would have required little or no irrigation. The naked varieties, although they required more water, were much easier to remove from the chaff after harvest. Therefore, periods in European prehistory when naked forms of barley were preferred may indicate that barley was being grown as a source of flour.

Interestingly, this preference for naked barley seems to have been reversed during the first millennium BC and the Roman period.[4] At first this sweeping trend would seem at odds with the sociopolitical developments of the time, notably the expansion of the Roman Empire, centralized public works projects (including the construction of aqueducts and irrigation systems), the introduction of crop-rotation cycles, and the development of new harvesting and processing tools, all of which led to significant population growth. One possible explanation is that irrigation systems allowed a greater portion of the population to switch to eating wheat breads and thus encouraged the cultivation of bread wheats. As these supplanted naked barley, the hulled form once again became a popular crop for fodder and fermentation. As irrigation increased in the first millennium BC, this same trend of a switch from naked to hulled forms of barley appears to have occurred across Central Asia as well.

THE ORIGINS OF BARLEY

The debate over the origins and spread of barley is far more complicated and involves a much larger corpus of data than the broomcorn millet debate. Domesticated barley can be traced back more than ten thousand years to the Fertile Crescent, where the grass appears to have been domesticated over several millennia.[5] The domesticated morphotype originated from a

Figure 9. A bread vendor polishing fresh loaves of flatbread, made in Uzbek style in the *tandir* oven, Bukhara bazaar, 2017. Photo by the author.

wild, brittle-rachised, two-rowed, hulled form (*H. vulgare* ssp. *spontaneum*).[6] However, the abundance of genetic data now available for *Hordeum* has stirred up considerable discussion.

Many researchers support the idea of two separate domestications of barley. Some have suggested, based on genetic data, that there were two centers of domestication, one in the Fertile Crescent and one somewhere farther east.[7] This idea has appealed to many archaeologists and historians; notably, researchers working in Tibet, where the cuisine is heavily dependent on barley, have used it to argue that barley was domesticated there. Additionally, David Harris postulates, in his book on early farmers of western Turkmenistan, that barley was independently domesticated in southern Central Asia, possibly in the ancient city of Jeitun or a neighboring related community.[8] A few geneticists who have attempted to grapple with the archaeological evidence have proposed that domestication could have occurred at either Jeitun or the ancient city of Mehrgarh, just west of the Indus Valley region of Pakistan. They note that Mehrgarh lies on the eastern edge of the range for wild barley and would have been an ideal site for domestication.[9]

While these theories are enticing, there are absolutely no archaeological data to support such suppositions. The earliest farming communities in southern Central Asia practiced a fully developed productive economy. The barley grown around Jeitun nearly eight thousand years ago was unquestionably domesticated, with large, plump grains and tough rachises, and in some cases it also expressed the dehulled (naked) mutation. Although the cultivation of barley at Mehrgarh and Jeitun may have contributed to the overall process of domestication and to the genetic makeup of modern Eurasian barley, the first settlers to establish these villages clearly brought with them fully domesticated barley from the Iranian Plateau.

Support for the idea that *Hordeum* was domesticated twice is largely predicated on the argument that two genetically distinct clades appear in the modern and historical populations of barley from Europe and Asia. The distinction is based on the fact that there are two alleles for nonbrittle rachises (the small stem that holds the grain to the ear).[10] The phenotype for the nonbrittle rachis is programmed by a mutation in one of two tightly linked genes, *Bt1* and *Bt2*. Geneticists generally consider the number of genetically unique mutations or alleles that code for a specific domestication trait to be equal to the number of independent domestication events for that crop.[11] This simplifying rule of thumb is valid for several cereal crops, which often have a single allele for nonbrittle rachises, but it has led to misinterpretations of the population genetics of many modern crops. With barley, division of the crop into two clades rests on the speculation that these two mutations for brittle rachises occurred independently and in isolation: in other words, neither mutation occurred in a population that already contained the other mutation or introgressed into a population from a wild source. Although it is not exactly clear what these two mutations imply, it is fascinating that many barley cultivars possess both mutations.

Other geneticists support a single domestication event for *H. vulgare*.[12] One team bases their argument for monophyly on more detailed genetic studies, genome-wide surveys, and multilocus systems, as opposed to a study of a specific targeted allele.[13] The idea of a single domestication event or gradual evolutionary process, presumably in the Fertile Crescent, is more consistent with the archaeobotanical data. However, the debate over a one- or two-center model is actually just the tip of the iceberg. The last couple of decades of genetic work on barley has raised more questions than it has answered. Geneticists have proposed centers of domestication as far afield as Morocco, Ethiopia, the western Mediterranean, and Tibet.[14] Some recognize two genetic populations and claim that there were two or more domestication events but do not specify where.[15]

Because barley is central to Tibetan cuisine, several scholars have sought to pinpoint a separate center of domestication for barley in Tibet.[16] Modern Tibetan barley is specifically adapted to short growing seasons, heavy frosts, and high elevations; however, there is currently no evidence to suggest that the original domestication genes, notably for nonbrittle rachises, evolved independently in this region. Nonetheless, the idea of an independent domestication is supported by the identification and classification of what are often referred to as truly wild populations of barley on the Tibetan Plateau. This purported wild barley has six rows of grains on each ear, like many domesticated forms of barley, whereas all other wild forms of barley have two rows of grains. Understandably, some scholars have questioned whether it could instead be feral barley—a domesticated barley that has escaped cultivation and now grows wild.[17] Regardless, subsequent genetic research into this question does not support the theory that these wild (or feral) barley populations are the progenitors of domesticated Tibetan barley.[18]

A more comprehensive genetic study that included both wild and domesticated populations of high-elevation barley has proposed an interesting and more likely theory: that domesticated barley, with tough rachises and large grains, spread onto the Tibetan Plateau, but genes from wild relatives introgressed into the gene pool of the cultivated population after its introduction.[19] This does not support the idea of a separate center of domestication, but it does suggest that early Tibetan farmers played a role, albeit an inadvertent one, in developing the cultivars that their descendants still harvest today.

In the end, the long-held view that barley was domesticated in the Fertile Crescent ten thousand years ago seems to hold up. The oldest archaeobotanical remains of cultivated barley that clearly show evidence for domestication come from the Fertile Crescent, and all proposals of other centers for barley domestication have failed to offer supporting data. Likewise, most of the recent studies of barley genetics point to southwest Asia, eliminating other proposed locations, such as Tibet. Despite decades of debate and dozens of genetic studies proposing a range of models, the macrobotanical data were correct all along. As some scholars have pointed out, the models that seem to closely reflect reality in plant domestication tend to rely on a combination of methodological approaches; models based on one specialized approach, be it genetics, microbotanical studies, or macrobotanical analysis, rarely hold up to close scrutiny.[20] However, new macrobotanical parameters for identifying traits of domestication and a more detailed analysis of the archaeobotanical data reveal that the history of barley domestication in the Fertile Crescent is more complicated than was once thought.[21]

The Fertile Crescent is the home of many of the crops grown across Europe and Asia in antiquity. Much was learned about the origins of agriculture here from studies conducted from the 1950s through the 1980s. However, the region comprises parts of several modern nations, and the issues of which areas should be studied and which research teams should be allowed access to archaeological sites have been subject to political turmoil. Therefore, our view of agriculture in the Fertile Crescent is actually rather fragmentary. Some areas, like Turkey and Israel, have been studied in detail, while others, including the eastern reaches abutting Iran, are blank slates.

French and German archaeological teams working in southwest and Central Asia have historically prospered from amicable political relationships with nations in this part of the world, at least compared to American research teams. In 2009 and 2010, a team of archaeologists from the University of Tübingen, in collaboration with researchers from the Iranian Center for Archaeology, excavated at Chogha Golan, a site at 485 meters elevation in the Zagros foothills in the Ilam Province of Iran, on the far eastern edge of the Crescent. The three-hectare site was occupied from 12,000 to 9800 BC. Wild barley was a significant food at the site, likely cultivated or maintained for much of the period of its occupation: the earliest layers at the site provided archaeobotanical evidence for early gathering of *Hordeum vulgare* ssp. *spontaneum*, as well as a variety of other wild progenitor crops.[22] The archaeologists noted an increase in the percentage of tough-rachised forms of triticoid grains (wild wheat relatives) by the later periods of occupation at the site, illustrating plant domestication on the eastern end of the Fertile Crescent roughly ten thousand years ago. At essentially the same time, a similar process of domestication was taking place on the western end, as attested by similar data sets from other sites in Iraq and Syria.[23] These parallel data indicate that domestication of the founder crops occurred simultaneously at different sites across the Fertile Crescent.

Following the Chogha Golan discovery, archaeobotanists have identified five clusters of sites across the Fertile Crescent where members of the founder crop complex were gradually being domesticated in unison.[24] They have identified roughly eleven distinct sites that provide indications for the domestication process. In addition, they have shown that the farmers in each of these five clusters were growing different sets of mixed crops, and that the specific varieties of individual crops were distinct from each other. This evidence seems to suggest that domestication of these founder crops occurred in each of these centers independently and in parallel, rather than at only one spot or at one point in time. This view of the Fertile Crescent as a site of multiple centers of domestication, likely interacting and also maintaining genetically distinct crop complexes, is ushering in a new paradigm in the debate over the origins of agriculture. Scholars are leaving behind the ideas of centers of domestication and rapid domestication events and instead looking for regions where crops were domesticated through long-drawn-out processes, likely involving complex genetic events.[25]

In light of this new paradigm and in response to the growing macrobotanical evidence for multiple parallel simultaneous domestications of barley within the Fertile Crescent, a new genetics-based model has been proposed that effectively refutes dozens of previously published genetic studies of barley domestication. This new approach isolates three distinct populations of barley in Europe and argues that the crops may have had distinct origins, *all* within the Fertile Crescent.[26] Scientists have also proposed that in the fifth and fourth millennia BC, a population of barley that was better adapted to northern latitudes spread into Europe from southwest Asia, and that the barley grown in southeastern Europe at that time may have spread into that region by a different route than the other two barley populations.[27]

If we accept that all domesticated barley lineages can be traced back to the Fertile Crescent of southwest Asia, we can break the process of barley domestication into four stages. The story of barley, like that of wheat, peas, lentils, and several other familiar crops, starts in southwest Asia. For millennia, early hunter-gatherers along the Tigris and Euphrates Rivers or in the nearby flanks of the Zagros Mountains collected seeds from wild barley plants. Through this coevolutionary relationship, the wild barley evolved to take full advantage of the human dispersal mechanism. The first trait indicative of domestication, a nonbrittle rachis, appeared in the grass population around 8000 BC, showing that human harvesting practices were by then applying enough artificial selective pressure on wild barley stands that they significantly altered the physiology of the plants.

By 6500 BC, domesticated barley was undergoing a second major stage in its artificial selection. In its wild state, an ear of barley has only two rows of seeds; in many modern landraces there are six rows on one ear. The six-rowed morphotype of barley arises from a mutation of the *Vrs 1* allele. This mutation may have originated repeatedly in different geographic areas at different times.[28] This trait clearly benefits the farmer, but it also indirectly benefits the plant: although many of its potential offspring are consumed by humans, it is more favored by farmers, and thus the plant is propagated more widely. This is the most appealing and basic way that plants "seduce" humans. Once crops are domesticated, people select for and replant seeds from the plants that provide the largest yield of food, whether that means more seeds per plant or larger fruit.

The next stage in barley domestication was the development of a thin, easily shed hull, leading to a form now known as naked barley. In the wild, barley seeds are protected by thick glumes or hulls, which protect the grain from pests and dehydration. However, these hulls have to be removed before people can eat the grain, a difficult and labor-intensive process. By 6000 BC a unique genetic mutation of the *nud* locus resulted in the naked phenotype of barley. Based on genetic evidence, scientists have suggested that this mutation is monophyletic.[29] Naked barley spread quickly, as it was a preferred food in harsh and high-elevation environments where wheat and millet cannot grow.[30]

The fourth and last stage in the domestication of barley was the loss of photoperiod sensitivity. This is a response to the change in day and night length during the growing season. Shorter periods of darkness are the signal to many plants, including wheat and barley in their wild state, that spring is approaching and it is time to flower.[31] However, the barley and wheat that these early farmers were cultivating had a winter habit, meaning that the crops were sown in autumn, cultivated through the winter to take advantage of the winter rains in the semiarid regions of the Fertile Crescent, and harvested in the early summer. The genes that coded for the plant to respond to changes in day length had to be bred out of the plant before it could grow in more northerly regions—although this was likely not a conscious choice by the farmers. As farming pioneers brought the plants to more northerly latitudes along the pre–Silk Road routes, their photoperiodism became confused by the shorter winter days and longer nights.[32] Northern-latitude winters are not conducive to plant growth. Therefore, these farming pioneers needed new varieties of barley (and wheat) that did not express photoperiodism.

Several millennia elapsed before farming moved into northern Europe along the routes that eventually became the Silk Road. One argument used to explain this delay is that existing crops were not viable in these northern regions and higher elevations. According to this theory, generations of farmers gradually bred a cold-tolerant form of barley that did not respond to changes in day length. The mutation may have existed in a few plants in the field; as farmers moved their crops to more northerly regions or higher elevations, their crop yields would have decreased. The plants that did not set seed did not spread their genes to the next generation; therefore, farmers may have initially experienced lower crop yields, but eventually these yields increased as more plants in the field had reduced photoperiodism through gradual evolution and selection. This novel phenotype was able to take on a summer habit, adapted to being sown in the spring and harvested in the fall.[33] This specific mutation, likely in conjunction with other traits, allowed barley and wheat to become two of the world's most important and widespread crop plants.[34]

The mutation was a response to the shutting down of the *Ppd-H1* allele, which controls the photoperiod response. One theory suggests that a wild population of the grass with this mutation already existed in the mountains of Iran, and domesticated barley from further east in the Fertile Crescent was inadvertently crossed with the wild line as farmers moved into the region. An alternative theory posits that a secondary domestication of barley, on the eastern edge of the Crescent, may have incorporated some of the members of this gene pool.[35]

This second monophyletic clade of domesticated barley comprises the spring barleys, which spread north into Central Asia and ultimately to China and into the Himalaya and Pamir mountain ranges. However, we know that spring forms of barley and wheat spread into China along northern routes of the Silk Road, and that winter forms of wheat are widely grown across southern China today. These facts imply either that after farmers spread the crop across northern China, the photoperiod sensitivity trait was turned back on in some populations, or that wheat and barley were also introduced into China by other routes, possibly through India or routes to the south.

BARLEY ON THE SILK ROAD

The earliest remains of a farming culture thus far identified in Central Asia are from the Late Neolithic and Chalcolithic village of Jeitun. The earliest immigrant settlers on the southern periphery of the Kara-Kum Desert brought with them the varieties of wheat and barley that they had grown in their homelands to the south, on the Iranian Plateau, including both naked and hulled barley. As noted above, the *nud* mutation for naked barley is monophyletic, and therefore we can envision these two lines being grown and cultivated independently in different fields: naked barley was planted separately from the hulled forms so that the crops would not crossbreed and produce seeds with a mix of both traits. Interestingly, the earliest settlers to move into Western Europe brought with them both naked and hulled forms of barley as well.[36] Archaeobotanical studies at other early village sites across southern Central Asia, dating from the sixth through the third millennium BC, have recovered a mixture of hulled and naked forms of barley: they include Anau North in Turkmenistan, Shortughai in Afghanistan, and Sarazm in Tajikistan.[37] Similarly, hulled and naked forms dating to this period are found at sites across the Iranian Plateau, including layers dating to the fourth millennium BC at the site of Godin Tepe in Iran.[38]

The history of barley in southern Central Asia gets more complicated over time. The data so far suggest that the shifting preferences for naked barley in Europe are mirrored in Central Asia (though further research may demonstrate otherwise). The transition to naked barley is not as clear-cut as the trend in Europe, however; several assemblages still express a mix of the two forms of barley.[39] Preserved grains of both hulled and naked barley have been found at Anau South, a few hundred meters from Anau North and Namazga V and VI (ca. 2500 BC) in southern Turkmenistan, along the edge of the Kopet Dag Mountains.[40] Both morphotypes were still grown and consumed by people in the same village or farmstead as late as ca. 1600 BC at locations such as sites 1685 and 1681 in Turkmenistan.[41] However, in assemblages from the second millennium BC at Tasbas in Kazakhstan, the dominant grain is naked barley, and there are no specimens that express the typical morphological features of hulled barley.[42] Likewise, based on published photos of several dozen barley grains from the late third and early second millennium BC site of Gonur Depe, it appears that naked forms were preferred.[43] The same is true for the nearby site of Ojakly in the Murghab delta of Turkmenistan.[44] Only a handful of full archaeobotanical assemblages from Central Asia date to between the fifth and second millennia BC, but the evidence from those sites suggests that naked forms were favored.

The hulled forms of barley started regaining popularity in Europe in the first millennium BC. The few archaeobotanical assemblages from Central Asia for this period are almost completely dominated by hulled barley. The best-studied first-millennium BC village in Central Asia is Tuzusai (410–150 BC; see chapter 5). At Tuzusai, barley is the most prevalent grain, followed by broomcorn millet and free-threshing wheat. Almost all of the barley is a large-grained hulled form.[45] In addition, hulled six-rowed grains and rachises of barley were recovered from the sixth- through fourth-century BC village of Kyzyltepa, in the Surkhandarya region of southern Uzbekistan.[46] While most of the barley in archaeobotanical assemblages from East Asia is naked, a few examples of hulled forms have been found at later sites, for example in the later occupation phases at Kyung-lung Mesa (AD 694–880) on the southern edge of the Himalayan Plateau near Nepal,[47] and at Mebrak and Phudzeling (1000 BC to AD 100) in Nepal.[48]

COMPACT BARLEY

The barley grains recovered from sites across Central and East Asia from the third and second millennia BC are short and rounded in shape. Although no proper systematic study of the dimensions of these grains has yet been conducted, the distinct morphology has been noted in several archaeobotanical studies. It is of interest because many of the grains from sites in Asia are shorter and smaller than contemporaneous grains found around the Old World.[49] Naked barley grains with this compact, semispherical structure have been identified at the ancient city of Sarazm;[50] Ojakly;[51] Miri Qalat, Makran;[52] several sites in Pakistan (e.g., Mehrgarh and Nausharo);[53] and Tasbas.[54] Most of the early examples of naked barley grains recovered in archaeobotanical studies in western China (before the first millennium BC) are similar in shape.[55] Naomi Miller has noted that the compact morphotypes of barley at the site of Anau are plumper than those found at Erbaba in Turkey.[56]

It is unclear what was going on genetically to cause this specific morphotype in ancient Asian barley and whether it represents a single genetic mutation or a case of parallel evolution. Some preferential traits associated with the compact size could have caused farmers to select for this kind of barley. There is a similarly compact form of wheat, which dates to the same period in the same parts of Asia, and in the next chapter, I discuss some theories about why these compact forms of wheat and barley may have been selected for. However, if this compact barley originated from a single gene pool that spread across Asia, it would support the idea that barley originally spread into China, along with the highly compact wheat, along one of the routes that later formed the Silk Road. Highly compact forms of barley dating to the fourth millennium BC have been found in Pakistan and parts of southwest Asia, to the third millennium BC in southern Central Asia, and to the second millennium BC in northern Central Asia and northwestern China. Although simply connecting the dots does not always hold up as an explanation for the spread of crops, for the sake of discussion we can simplify the situation and envision the earliest barley being cultivated in the valleys along what would become the Silk Road, which originated from the same genetic population of barley that eventually colonized East Asia. This population had a short, compact stature; it was hardy, frost tolerant, and non-photoperiod responsive; it had a naked phenotype; and it was a six-rowed form. This is not to discredit the theory that a secondary wave of domesticated barley spread into southern China, along with winter wheat, through the southern Himalaya—possibly through the Pamir Mountains and maybe the Swat Valley. In fact, these morphological observations lend support to the idea of a secondary spread of barley and wheat into southern China, which replaced the earlier, compact forms.

BARLEY ON THE ROOFTOP OF THE WORLD

While barley may no longer be the preferred flour for bread making in Europe, it is still the main grain on the rooftop of the world, the Himalayan Plateau. Naked barley was prominent at high elevations across greater Central Asia before colonial expansion. When the British explorer Alexander Burnes passed through the high peaks of the Pamir and Hindu Kush in 1832, he noted that the local populations grew "a barley in this elevated country, which has no husk, and grows like wheat, but it is barley."[57] Describing farming in the Hindu Kush, he noted that a variety of fruits and nuts were grown in the river valleys and that few other crops survived at high elevations.

Few paths of the Silk Road ever ran over the Himalayan Plateau, although a recent discovery of tea leaves (*Camellia sinensis*) at the high-elevation city of Ngari (Ali) in western Tibet has been used to argue that tea was moving over the high peaks from Chang'an (modern Xi'an) as far back as AD 200.[58] The residents of this region played important roles in Silk Road trade and the development of Central Asian cuisines, especially during the later first millennium AD. In addition, the southerly foothills and valleys of the plateau may have been an important route for the early spread of southwest Asian crops into East Asia and of East Asian crops into South Asia, notably the Indus Valley.[59]

The cuisine of Tibetans, and other closely related ethnic groups in the modern regions of western China, Nepal, and northern India, is rather removed from the culinary traditions of the rest of Asia and those of Europe. Naked barley is at the center of nearly every meal. The staple of Tibetan cuisine is barley cakes (*gyabrag*) made with barley flour, yak butter, dry cheese curds, and sugar. The barley cakes are also used as ritual offerings to local spirits or Buddhist deities. Flour made from roasted barley groats, referred to as *tsampa*, is also used to make Tibetan tea and breads.

Because of barley's tolerance of frost and high altitude, it remains a key component of Tibetan farming. Yak herders cultivate naked barley in fields at elevations up to 4,500 meters, growing a six-rowed naked form specifically adapted to this harsh ecology. It is planted around April and harvested in August or September, before the arrival of the bitter winter nights. Tibetans grow this unique crop in some of the most unlikely soils, on the edges of hills and in dry fields that are frozen for most of the year.

The origins and spread of barley, and agriculture in general, through the Himalayan Plateau is a question of great interest to social scientists. Understanding the development of Tibetan culture requires an understanding of its dependence on barley. Moreover, understanding how agriculture was developed in one of the harshest ecological regions on earth enhances our understanding of human adaptability in general. In recent years, there has been an explosion of new archaeological research on the Himalayan Plateau, which is rapidly changing the scientific view of human adaptation.[60] Himalayan populations are genetically adapted to the cold, hypoxic environment in several distinct ways that are the result of hundreds of generations of selective pressure.[61]

While hunting bands may occasionally have made their way up to these elevations as early as twenty thousand years ago (or possibly as long as eighty thousand years ago, according to one view),[62] the first farmers to populate the higher regions arrived from the east around 5000 BC.[63] The process took millennia, and many would have perished through stillbirths, extremely high infant mortality, frostbite, and health complications associated with the low-oxygen and arduous environment.

Farther west, hunter-gatherer groups seasonally headed up the Pamir Mountains to hunt game in Tajikistan during the Paleolithic. The remains of a hunting camp at Osh-khona in the eastern Pamir, at 4,000 meters elevation, date to the Late Paleolithic. The camp consisted of an open-air fireplace where wood charcoal was burned, suggesting that wood resources may have been more abundant at these elevations in the past. Artifacts from the site also show evidence of craft production, including awls for leatherwork and stone flakes from napping. Abundant bones of small and large mammals and an arrow preserved with its shaft intact provide evidence of hunting.[64]

Eventually, these early visitors and denizens adapted to the high-altitude regions, breeding lines of crops and fostering descendants who could survive on the rooftop of the world. These descendants evolved positive haplotypes of the *EGLN1* and *PPARA* genes, which changed their hemoglobin phenotype. Hemoglobin is the component of blood that carries oxygen to the cells; in most humans, this system is not adapted to function well at higher elevations, where less oxygen is available. Modern Tibetans have a better oxygen-transport system (which in fact involves lower levels of oxygen-carrying hemoglobin in the blood) that evolved over thousands of years. This mutation is a unique response to hypoxia seen only among Tibetans.[65]

Interestingly, the earliest farmers to move into these higher elevations did not know about barley. Some linguists have suggested that the Tibeto-Burman (Sino-Tibetan) language phylum initially spread from the Sichuan Plains, or somewhere just to the north, in concert with the spread of millet-based farming.[66] Although it has received considerable criticism, many scholars still favor this explanation for the spread of sedentary farming people into higher elevations of the Himalaya.[67]

Millet-based agriculture is likely one of the cultural and physiological adaptations that allowed people to settle the eastern foothills of the Himalayan Plateau in the fourth millennium BC.[68] Millet is a highly adaptive crop that requires only a very short frost-free growing season and thus was well suited to cultivation at higher elevations where rice could not grow. However, it was eventually replaced by high-elevation-adapted barley, which is even better suited to the mountain environment. With the establishment of farming and yak herding, people started setting up year-round villages at these higher elevations. One team of scholars has provided a compelling argument that yaks may have been a key to the successful colonization of the Himalayan Plateau by sedentary people (although this colonization process started at least a millennium before there is clear evidence for yak domestication).[69] This argument is based on the use of yaks not as a source of meat or protein, but rather as a source of fuel for heating and cooking in an environment lacking in woody vegetation. Yaks produce nice, compact cakes of dung that provide a slow-burning, renewable fuel.

The introduction of wheat, barley, and millet, along with sheep and goats, helped settle this last ecological frontier: but, as at least one scholar has pointed out, these occurred too late to make them the key elements of

Himalayan settlement. Notably, cultivated crops were known to the people on the eastern Himalayan Plateau by at least 4,500 years ago, but the expansion of farming communities into this region did not take place until roughly 3,600 years ago. It is more plausible to ascribe colonization of the area to a suite of cultural and physiological adaptations, not to farming and pastoralism alone. The human physiological adaptations to hypoxia seem far more important than farming to establishing a sustainable high-altitude population: growing food is futile if birth complications and child mortality consistently deplete the population. The adaptation of crops to high elevations, in concert with this suite of human adaptations, is a wonderful example of the coevolution of culture and genes.[70]

The archaeological site of Karuo (map 3) is often referenced as the earliest sedentary farming community to have been established at higher elevations in western China, or at least one of the best-excavated examples. It is located on the bank of the Mekong River near the city of Qamdo (Changdu), in the Karub District of the Tibetan Autonomous District, at an elevation of 3,100 meters.[71] The site, first excavated in the 1970s, was revisited in 2002.[72] Radiocarbon dating of organic remains at the site has been problematic: the dates range from 3966 to 2196 BC, in three phases, two of which are inverted, (with sediments from different archaeological layers intermixed), but a targeted dating of domesticated grains has provided a narrower range, ca. 2700–2300 BC.[73]

The earliest settlers at Karuo seem to have been sedentary, living in small, semisubterranean houses.[74] They relied on hunting, fishing, and gathering but also raised pigs and millet, both broomcorn and foxtail.[75] Among the wild species that were hunted by the occupants, as attested by the zooarchaeological study, were goat, bovids, pig, red deer, antelope, hare, and macaque. Wild foraged plants included raspberries (*Rubus* sp.).

Building on these preliminary findings from Karuo, a much larger and more detailed study by an international team of scholars was recently

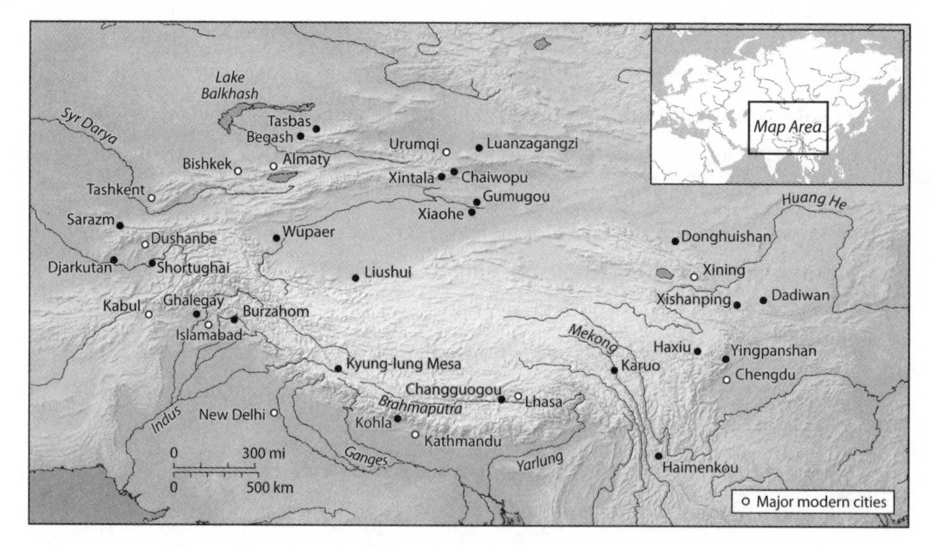

Map 3. Key archaeological sites in the high-altitude regions of Central Eurasia.

published. This study deals with archaeobotanical remains from fifty-three archaeological sites dating from 3200 to 300 BC across the Himalayan Plateau. The conclusions support the idea that the earliest agriculture in Tibet was based on the two millets. The data consist of archaeobotanical material from twenty-five archaeological sites dating to between 3200 and 1600 BC. Remains of the two millets were found at all these sites, but none of these is above 2,527 meters elevation. Of the remaining twenty-nine archaeological sites in the study, dating between 1600 and 300 BC, seventeen are located at elevations above 2,500 meters.[76] These findings suggest that although millet was an early crop in the region, its cultivation was restricted to rich, protected valleys and lowland regions. Not until specific breeds of high-elevation barley were developed were people able to settle the plateau.

A higher-elevation and more recent settlement is Changguogou in Gongkar County, west of the city of Lhasa (map 3). Changguogou lies on the

Yarlung Tsangpo River (the Tsangpo-Brahmaputra) in southern Tibet, at 3,600 meters. The settlement, which was occupied from 1400 to 800 BC, represents a well-established agropastoral community, relying on a range of domesticated plants as well as sheep and goat. The plants reported at the site include both types of millet, naked barley, wheat, rye (*Secale cereale*), possibly oats (most likely *Avena*), and a single pea.[77] It is possible that rye and oats made their way to this area along with other southwest Asian crops, but photos were not published of either grain, and the finds might represent distorted barley grains, an easy misidentification. Rye and oats are historically documented as being grown by various ethnic groups in the region, and preserved oat grains were found at the site of Kohla in the highlands of Nepal (AD 500–1500), supporting the identifications at Changguogou.[78] A handful of wild foraged plants were also identified at Changguogou, including a pine-nut shell and a whole preserved root of drolma (*Argentina anserine*), a rhizome still eaten in Tibet today.[79]

The international team of archaeologists and archaeobotanists that worked at Karuo also studied botanical remains from twenty-nine sites at a range of elevations that were roughly contemporaneous with Changguogou.[80] These sites contained both types of millet as well as wheat and barley. Sites at higher elevations were dominated by barley, with occasional finds of broomcorn millet and wheat. Sheep and goat bones were very common at sites above 3,000 meters. The data make it clear that by this period, a high-elevation agropastoral economy was taking shape across the previously unsettled frontiers of the Himalayan Plateau.

Excavations of funerary caves at Mebrak (3,500 meters) and Phudzeling (3,000 meters), in the Jhong Valley of the formerly isolated and restricted region of Upper Mustang in Nepal, show that the sites date from 1000 BC until modern times.[81] The archaeobotanical material recovered from these excavations, collected between 1990 and 1995, provides a wonderful depiction of the development of high-elevation cuisine, as well as the spread of southwest Asian crops eastward along the southern edges of the Himalaya. The archaeobotanist for this project suggested that people in the region were maintaining cattle herds and farming buckwheat, identifying what was labeled as *Fagopyrum* cf. *esculentum*, *F. tataricum*, and both naked and hulled barley dating to between 1000 and 100 BC.[82] He also identified another interesting southwest Asian domesticate, flax (*Linum usitatissimum*). The study noted finds of rice, lentils (*Lens culinaris*), hemp, apricots, and rosehips (*Rosa*) in contexts dated to later than 400 BC.

The history of the introduction of southwest Asian crops to the Himalayan Plateau and the high elevations of Central Eurasia is complicated and not well understood. However, archaeologists have shown that there were close cultural ties between populations across Tibet, Qinghai, and Xinjiang starting in the second millennium BC. Botanical finds at Ghalegay, in the Swat region of Pakistan, from layers as old as 1900 BC, clearly illustrate that certain southwest Asian crops were moving along the southern foothills of the Himalaya by the second millennium BC. Supporting this reasoning, wheat, barley, peas, and lentils have been found at Burzahom, Gufkral, and Semthan (ca. 2800–2300 BC) in Kashmir.[83] Clearly, by about 2500 BC, similar economic processes and the same crop assemblage were in place across the mountain foothill zones of Eurasia, from the Kopet Dag Mountains to Kashmir. The analysis of material culture from grave goods at the cemetery site of Liushui (1108–893 BC) in the Kunlun Mountains has also identified close stylistic similarities between people on the Himalayan Plateau and those of the Central Asian steppe, starting in the second millennium BC.[84]

The transition to barley-centered farming in Tibet seems to have taken place sometime during the latter half of the first millennium BC. Kyunglung Mesa (AD 220–334 and 694–880), situated on a rocky outcropping at 4,250 meters, once housed a large agropastoral community. The archaeobotanical remains recovered from the site include abundant barley grains and a couple of wheat grains (and rachises of both crops).[85] New archaeobotani-

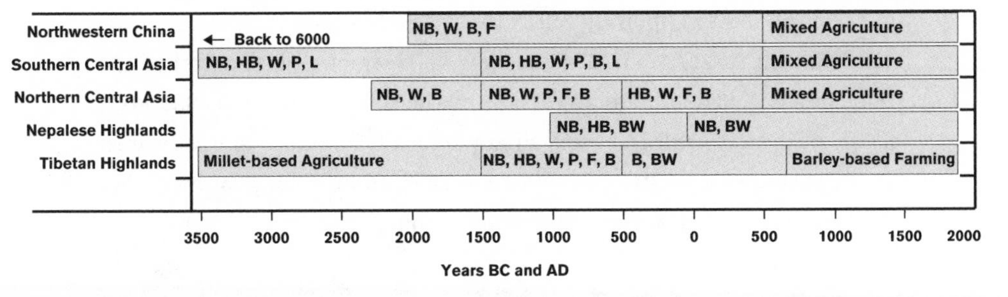

Figure 10. Agricultural development across Central Asia, by major regions. Data from d'Alpoim Guedes et al. (2014); Knörzer (2000); and Spengler, Frachetti, et al. (2014). B: broomcorn millet. BW: buckwheat. F: foxtail millet. HB: hulled barley. NB: naked barley. L: lentils. P: peas. W: wheat.

cal finds from other regions in Tibet provide evidence of similar transitions to farming. For example, seven archaeological sites from the late Holocene in the Nujiang River Valley, in southeastern Tibet, provide evidence that hunter-gatherers lived in the mountain valleys on the southeastern margins of the plateau before 5800 BC, but that millet farmers did not settle in the region until around 2200 BC.[86]

By the middle of the first millennium AD, barley had become central to the culture and identity of peoples in the high-elevation Himalayan region, as attested by literary sources from dynastic China as well as by finds of the grain in burials.[87] The success of this barley-based farming system is evident in the expansion of the Tibetan Empire (AD 618–842). The surplus of grain made these agropastoral yak herders on the rooftop of the world so powerful that they were able to knock down the gates of Chang'an (modern Xi'an), the Tang capital of China, in 763 and gain control of the southern Tarim Basin and key exchange routes. The Tang struggled to hold their sources of foreign merchandise and the routes of the northern Tarim Basin, a route for the import of exotic goods. From roughly 600 to 866, Tibetans held noncontinuous dominion over important Silk Road routes, but they were in continual conflict with Turks, Arabs, and the Tang.[88] The empire that controlled the trade routes held dominion over Asia, and the key to holding the Silk Road lay in grain stores sufficient to feed hundreds of thousands of cavalrymen and their horses. The struggles between Tibet and the Tang for control of these trade routes reflect shifts in resource availability and crop production.

CONCLUSIONS

The story of humanity's coevolution with the small, unobtrusive *Hordeum* grass began well over ten thousand years ago, in the Fertile Crescent of southwest Asia. Groups of foraging women collected seeds from seasonally ripening wild patches of *H. vulgare* var. *spontaneum*—a two-rowed, brittle-rachised, hulled form of wild barley. The plant's relationship with these people caused it to evolve in ways that made it more attractive to its human companions. By 6500 BC, the first farming pioneers reached Western Europe, and by 6000 BC, farming immigrants arrived in southern Central Asia. All of these early immigrants carried with them pouches of *Hordeum* seeds that were already dramatically different from their wild counterparts. The barley grains came in two morphotypes, hulled and naked; they had tough rachises and large, plump ears with six rows of grains. For thousands of years, farmers across Europe favored barley because of its low labor demands, and by the second millennium BC, the cultivation of naked forms predominated across all of Europe as well as in Central Asia, reducing the amount of processing necessary to turn the harvested grains into bread.

By the first millennium BC, with the introduction of government-regulated grain stores and irrigation projects, farmers began to phase out the naked barley in their fields in favor of free-threshing wheat. This form of wheat produced whiter, lighter bread, the preferred food of the affluent across Europe. Barley became the crop of the poor: its social and culinary roles changed, from bread to beer and from cherished to shunned. However, in Tibet, a form of naked barley, adapted to the harsh mountain environment, remains a staple of the diet and the culture. And elsewhere in the world, barley still retains an essential role as a principal ingredient of beer and whiskey.

CHAPTER SEVEN

The Wheats

Wheat is strongly associated with the bread-baking cuisines of West Asia, Europe, and North America; however, it is also an important crop in China. In modern times, wheat has become the second most heavily planted grain in China, after rice. By two thousand years ago it was gaining prominence in East Asian cuisine in the forms of noodles, dumplings, and steamed buns. The importance of wheat in what is now northern China resulted in significant changes to crop cultivation systems.

The Chinese word *mai* refers to both wheat and barley, along with the other large-grained cereals that originated in regions west of China. Wheat is known as lesser or *xiao mai*, and barley as greater or *da mai*. By the twelfth century, and likely much earlier, winter wheat was being grown in a rotation with a summer rice crop across much of China—especially in regions to the south of the millet-growing zones and to the north of the two-crop rice-growing zones of the far south. Historians claim that the intensification of agriculture in the Yangtze region through crop rotations was due in part to the fall of the Northern Song (960–1126). This led to waves of refugees emigrating south, bringing the knowledge of wheat cultivation with them. By the seventeenth century, as much as 50 percent of the crop harvest in the northern regions of China was based on winter wheat.[1] This chapter traces the paths that wheat took to reach this position of prominence in Chinese agriculture.

Just as millet allowed Mediterranean farmers to produce a summer crop on irrigated lands, winter wheat allowed Chinese farmers to grow a winter crop on either drained rice paddies or, in the north, on millet fields. As with millet in Europe, the sharp increase in grain supply that resulted from these crop rotations fueled rapid population growth and more investment in economic activities other than farming. This led to craft specialization and the rise of an educated scholarly class—not to mention the expansion of military forces, which amplified conflict.

Wet paddy rice agriculture is not, however, straightforwardly compatible with the cultivation of *mai*. Paddy fields had to be thoroughly drained in the fall before winter wheat could be planted; in many cases farmers mounded up soil or built ridges to plant the wheat on, and sometimes they even dug secondary drainage systems. While crop rotations massively increased yields, they incurred high labor costs, requiring farmers to toil in their fields throughout the winter and the summer.

ANCIENT VARIETIES OF WHEATS

Of all domesticated plants, the *Triticum* grasses have the strongest and most deeply intertwined coevolutionary bonds with humans. Early Natufian foragers started on the road to domesticating wheat as early as 9500 BC. These were the same foragers, or at least the same general cultural group, who brought barley under their dominion; hence, as with so many of the domesticated foods on our tables today, the narrative of the wheats began in the Fertile Crescent.[2] By 8500 BC, domesticated forms of wheat constituted a majority of the plants under cultivation. The route that the wheats took toward domestication was far from linear, however, and likely followed many simultaneous domestication pathways that merged and split, or in some cases reached dead ends.

What most people think of as wheat is thus actually a complex of species, each with its own history. This complexity is concealed by the use of one vernacular term to cover them all. By contrast, when we discuss the legumes, we do not refer to them all as beans; we recognize their differences by calling them lima beans (*Phaseolus lunatus*), green beans (*P. vulgaris*), chickpeas (*Cicer arietinum*), or grass peas (*Lathyrus sativus*). Although all the wheats are members of the genus *Triticum*, there are five genetically isolated species.

Because of their prominence in the Euro-American kitchen, the wheats have received the most archaeobotanical and phytogenetic attention of all the Old World crops; even so, the narrative of their original domestication and spread is still being sorted out. It is a fascinating and complex story, involving layers of hybridization (expressing polyploidy, or whole genome duplication), tens of thousands of years of isolation and then hybridization between species, and the struggles of generations of farmers over several millennia. This chapter can offer only a brief account.[3]

Understanding the clues that archaeobotanists have unraveled to trace the journey of the wheats along the Silk Road requires some knowledge of the unique genetic structure of these species, beginning with the concepts of hybridization and polyploidy. When members of two distantly related species of organisms mate and produce offspring, the process is referred to as hybridization. Their offspring can resemble either or both of the parents in different respects; in some cases, especially when the parents are very distantly related, it can look like a completely different organism. When the two parents are so different that their chromosomes cannot even match up correctly during reproduction, the chromosomes may be completely duplicated, so that the offspring has a full set of chromosomes from each of its parents rather than half a set from each. This phenomenon, which is common in plants and extremely rare in animals, is known as polyploidy. The polyploid offspring almost always looks different from the parents and can no longer reproduce with the population from which it originated: therefore it constitutes an entirely new species. This process occurred several times during wheat domestication, resulting in some species with two, four, or six sets of chromosomes, referred to as *diploid*, *tetraploid*, and *hexaploid* species, respectively.

Of the five genetically distinct forms of wheat, two are diploid: *Triticum monococcum* (einkorn wheat) and *T. urartu*. The *T. monococcum* wheats have wild forms that closely resemble their domesticated relative. *T. urartu* was never domesticated, and while closely resembling its wild diploid relative, it is genetically isolated. Einkorn was domesticated in southwest Asia by the late ninth millennium BC from the wild form *T. monococcum* ssp. *monococcum*. All the *monococcum* subspecies (einkorn) are closely related and hard to distinguish morphologically or genetically. It is now known that *T. urartu* donated its chromosomes to the polyploid complex that makes up the tetraploid and hexaploid wheats. This happened through natural gene-flow processes, long before human manipulation of the genus.

Hexaploid wheats include the modern, free-threshing bread wheat (*Triticum aestivum* ssp. *aestivum*); they contain high proportions of gluten, the protein that makes bread dough elastic and enables a loaf to rise and become light when baked. The tetraploid wheats, which are lower in gluten, are known as macaroni or durum wheats. They have played a significant role in European cuisine, notably as the main ingredient in dried pastas and hardtack breads. Hardtacks provide a nonperishable and easily stored source of carbohydrates but one that is generally regarded as less palatable than fresh bread.

There are two species of tetraploid wheats, *T. turgidum* and *T. timopheevi*. Molecular and cytogenetic research has shown that one of the genomes of both of these tetraploids originated in an *urartu*-like ancestral wheat.[4] *T. timopheevi* is an endemic domesticate, meaning that it originated in and is still mainly grown in a small area of Georgia, in the Caucasus, and is therefore not of relevance to this book. By contrast, *T. turgidum* (emmer wheat), spread across much of Eurasia and has been archaeologically identified at

Jeitun in southwestern Turkmenistan, on the edge of Central Asia.[5] *T. turgidum*, or emmer, was domesticated from a wild tetraploid, *T. turgidum* ssp. *dicoccoides*, in the ninth millennium BC, around the same time that einkorn wheat and hulled barley were domesticated in southwest Asia.

Like hulled barley, the earliest domesticated wheat and all of the wild wheat relatives have hulls or shells that protect the grains as they develop and disperse; humans need to remove these hulls before they can eat the grain. Early farmers recognized and selected for "free-threshing" mutations in domesticated wheat, which have paper-thin shells that fall off easily. Complicating the story of wheat domestication, there are free-threshing tetraploid as well as hexaploid wheats, the most prominent of the former being durum.[6]

Understanding the spread of free-threshing tetraploid wheats into Central Asia is important because the majority of wheats grown across the southern Islamic world today are durum wheats. One historian argues that they spread with the Arab conquests and the diffusion of Islam; this theory is consistent with some of the currently available data.[7] In addition to the typical macaroni wheats (*T. turgidum* ssp. *durum*), farmers in southern Central Asia and across the Iranian Plateau and Mesopotamia grow Khorasan wheat (*T. turgidum* ssp. *turanicum*), a variety historically attributed to northern Iran, and Persian wheat (*T. turgidum* ssp. *carthlicum*), a lesser-known morphotype grown across southwest Asia in the past. How far east from the Fertile Crescent the tetraploid wheats spread in antiquity is not fully known. The only clear evidence for the crop in Central Asia proper comes from sixth- and fifth-millennium BC layers at the archaeological site of Jeitun. The lack of any discoveries of *T. turgidum* in the heart of Central Asia suggests that it was likely replaced by hexaploid wheats in the fourth millennium BC. All of the archaeobotanical remains of wheat thus far reported in China are morphologically closer to the free-threshing hexaploid form *Tritium aestivum* ssp. *aestivum* (see below).[8] Archaeobotanists working in East Asia have acknowledged that early wheat in China (2600–1500 BC) and later wheat from Korea (ca. 1000 BC) and Japan (beginning of the first millennium AD) are hexaploid.[9] Genetic studies of remains of wheat grains from cemeteries in the Lopnor region of Xinjiang, combined with study of early herbarium material and historic records, have shown that all of the earliest examples of wheat in China are from a free-threshing hexaploid wheat.[10]

Village sites in Syria dating to the early Islamic period contain two different forms of free-threshing wheat, a form that has small grains and compact ears (compact wheat) and a form with long grains and long ears (lax-eared wheat). In these ancient villages, there are also limited remains of rachises from a free-threshing tetraploid wheat, although all the actual preserved grains from these sites contain only hexaploid free-threshing wheat grains; the only evidence of tetraploid wheats is in the form of rachises.[11] The only evidence thus far for durum-type wheat in Central Asia comes from Tashbulak, the high-elevation medieval town in Uzbekistan (see chapter 2). While working at the site, I recovered a few wheat rachises from Tashbulak that express the clear morphology of a free-threshing tetraploid.[12]

In Europe, diploid einkorn (*T. monococcum*) and tetraploid emmer wheat (*T. turgidum* ssp. *dicoccum*) remained important crops for millennia after the introduction of free-threshing hexaploid wheats.[13] Interestingly, despite being absent from East and Central Asia, the free-threshing tetraploid wheats, along with emmer wheat, became important in India.[14] As crops spread north from India into Kashmir, and northeast from the Iranian Plateau into Central Asia, the hulled wheats were left behind (although free-threshing tetraploid emmer wheat from the third millennium BC has been reported in Kashmir).[15] At this time in Kashmir, other southwest Asian crops—including emmer, peas, and lentils—were grown in concert with crops that were domesticated in India, such as black gram (*Vigna mungo*) and mung bean (*V. radiata*).[16] These latter crops—but not the hulled or free-threshing wheats—also made it south into India and onto the upper Punjab Plain by 2500 BC.[17]

The fifth wheat species in the *Triticum* complex is the hexaploid *T. aestivum*—common bread wheat. This species evolved under cultivation from a polyploid cross between a tetraploid *T. turgidum* (already containing the genome mentioned above from *urartu*) and a distantly related wild grass (*Aegilops tauschii*), providing a new genome. Hexaploid wheats include a range of varieties, broken into two groups: the hulled (glume) and free-threshing wheats. Hulled hexaploid wheats include *T. aestivum* ssp. *spelta* (spelt) and *T. aestivum* ssp. *macha*, which is endemic to western Georgia.[18] Free-threshing hexaploid wheats (the bread wheats) are easier to process after harvesting, and in many parts of Eurasia they replaced emmer or durum wheat as the preferred crop during the fourth or third millennium BC.

THE WHEAT ROAD

The questions of how, from where, and when wheat spread into China and eventually became a vital part of Chinese cuisine and farming have engaged scholars for over a decade.[19] Interest has grown in recent years,[20] following the elucidation of what Zhijin Zhao from the Chinese Academy of Social Sciences has called the "wheat road": a mountain corridor along which wheat diffused into China in the third millennium BC.[21]

Wheat has been identified, in very low abundance, at a number of sites dating back to the very end of the third millennium BC, with questionable earlier finds. Most of the early finds of wheat are extremely sparse compared to remains of millet and rice. Symbols on oracle bones that date to the Shang Dynasty (ca. 1558–1046 BC) have been interpreted as early forms of the characters for wheat and barley, suggesting that the crop was present in China by then.[22] Wheat is rare in early texts and oracle inscriptions from this period, though there are hundreds of mentions of millet.[23] Both wheat and barley are mentioned in the *Book of Odes* (*Shijing*), a compilation of poems dating from between the eleventh and seventh centuries BC.[24] It seems clear that wheat was relatively uncommon in the farming communities of northern China during the second millennium BC and that its cultivation probably did not become widespread until the Han Dynasty (206 BC–AD 220).[25] Its dramatic rise to prominence in that period may have been due in part to the introduction of crop-rotation cycles using winter wheat—varieties of wheat that are planted after the summer crops are harvested and grow through the winter, sometimes withstanding heavy snow cover, to be harvested in the late spring or early summer.

The fragmentary and limited nature of archaeobotanical finds of wheat makes it difficult to piece together the story of its spread across Asia. As archaeobotanical studies in China have increased, along with the number of rescue excavations mounted in advance of urban sprawl, more wheat remains have been found. These finds have prompted investigations into the routes of introduction of wheat and mechanisms of social change associated with its adoption into an agricultural system that already included indigenous grain crops.

As with many of the crops I discuss in this book, current data suggest that wheat spread into China along the foothills of the Central Asian mountains, essentially reversing the route by which millet spread out of China. The mountain rains and glacial-melt streams would have provided a rich environment where farmers could experiment with agriculture and eventually develop regionally suited varieties of wheat (and other crops). As with barley, the spread of cereals into northern latitudes may have been delayed by a need for specific phenotypes of the crops that had daylight-neutral genes and frost tolerance.

The earliest domesticated remains of free-threshing wheat identified north of the Pamir Mountains in Central Asia were found at one of the archaeological projects that I collaborated with in Tasbas, Kazakhstan, a small encampment or possibly a mobile pastoralist seasonal settlement dating to ca. 2600 BC.[26] I also identified the botanical remains of similar free-threshing wheat from the nearby site of Begash (ca. 2200 BC), where I found it in association with broomcorn millet and a single barley grain.[27] Begash also likely originated as a small seasonal encampment of mobile pastoralists. At both sites, the preserved carbonized grains were recovered from the earliest occupation phase, phase 1a, which is characterized by flagstone-lined cists (burial chambers) believed to hold human cremation remains.[28] The grains were interred inside a burial as mortuary offerings, but the ceremonial context in which they were found leaves open questions regarding their use in the domestic economy. Archaeobotanists are still searching for early hearth features or kitchen cooking areas from this period in eastern Central Asia. Flagstone or slab-burial cists were common in Central Asia from the third millennium BC through the end of the first millennium BC.[29] Other grain offerings have been recovered from burials across Asia, notably from Xinjiang in northern China and Mongolia.[30]

Free-threshing hexaploid wheat was established as part of the farming system in the central plains of China by the middle of the second millennium BC.[31] A number of published reports of earlier preserved wheat grains in China have been disputed.[32] Carbonized wheat grains were recovered from Liangchengzhen, Shangdong Province, in the Longshan culture (2600–1800 BC).[33] Other researchers have subsequently pointed out, however, that the two well-identified grains from that site are not directly dated, and neither are other Longshan grains, from Zhaojialai in Shaanxi Province and Baligang in Henan Province.[34] If these early dates for wheat on the central plains of China are rejected, it is easier to support the theory that wheat was introduced to China through the mountainous Hexi or Gansu Corridor.[35] Early finds of wheat in Kashmir and the southern Himalaya also leave open the possibility that wheat and barley farming spread across the entire Central Asian mountain zone like a wave rather than along a single corridor. The best attempt at a synthetic study of the distribution of these compact wheats, looking at a chronological spread across Asia, has suggested that grain size gradually decreased as the crops moved eastward.[36] The reasons for the distribution of this form of wheat are still disputed: some scholars link it to selection for varieties that are adapted to specific kinds of poor growing conditions, such as high elevation.

The idea that free-threshing hexaploid wheat entered China from the northwest periphery is supported by finds of desiccated wheat grains in burials from the late-second- and first-millennium BC cemeteries at Gumugou, Xiaohe, and Sampula,[37] as well as at Luanzagangzi,[38] all in Xinjiang. A more interesting find of early wheat comes from Xishanping in Gansu Province. Archaeologists working at the site claim that not only wheat but also barley and possibly oats were recovered from archaeobotanical samples dating to between ca. 2700 and 2350 BC.[39] The wheat from Xishanping is a lax-eared form, unlike most of the early Central Eurasian wheats. However, other scholars have questioned the dating at this site.[40] Many other grains from northwestern China have been securely dated and were preserved in funerary contexts. Grains are often recovered from burials across Inner Asia, either in containers or sprinkled over the body. Wheat grains were found among the grave goods recovered from burials in the Xiaohe cemetery dating to 2011–1464 BC,[41] and among the sediments from other sites in Xinjiang, notably Sidaoguo (1493–1129 BC), Xintala (2006–1622 BC), and Wupaer (1189–418 BC).[42] Archaeologists have presented a set of radiocarbon dates run directly on ancient grains of free-threshing wheat and naked barley grains from the site of Donguishan in the Siba culture, noting that the site was occupied between ca. 1550 and 1450 BC.[43] The sites of Huoshiliang (ca. 2135–1895 BC) and Ganggangwa (ca. 2026–1759 BC), from the same culture group, provide slightly earlier radiocarbon dates, also directly run on grains of wheat.[44]

In Xinjiang, wheat and barley grains were found sprinkled on the floor of tombs dating from the late second millennium BC all the way into the Tang period.[45] In the Turfan region of Xinjiang, a large number of astonishingly well-preserved tombs have been excavated from the desert sands. One of the oldest, with burials dating to about 1000 BC, is the Yanghai cemetery, where many of the graves contained wheat, barley, and broomcorn millet.

The burials are often contained in coffins made of wood, or branches woven to imitate wooden boards—presumably because large planks were an extremely precious commodity in a desert oasis. The coffins often contained pottery and other funerary offerings, such as leather objects, wool textiles, bronze tools, and sheep or horse meat, particularly sheep heads or shoulders. Burials at the nearby Jindian cemetery (200–50 BC) contain many of the same offerings, in addition to iron and stone artifacts such as agate and glass beads.

More recent burials are even more interesting. From the elaborate Tang Dynasty burials of the Astana cemetery, Chinese archaeologists have recovered bronze and iron tools, paintings, wooden artifacts, gold, silken textiles, and various foods. These included a kettle containing desiccated wheat dumplings, which resembled those eaten in the region today. Preserved wheat noodles were found at the nearby cemetery of Su Beixi.[46] The round, dense loaf found at the Astana site, which the Chinese scholars claim contained both barley and millet flour, is probably typical of the common food on the Silk Road for much of the past two millennia.

On the Himalayan Plateau at the site of Changguogou (1500 BC), remains of naked barley, free-threshing wheat, possible oats (*Avena*), and even a green pea were reported (see chapter 8).[47] These finds in the hills of western China may represent a second, more southerly spread of southwest Asian grains into China: more recent examples of these grains have been found in the south, and it is possible that by the second millennium BC or earlier they had dispersed so widely through the mountain foothill zone of Inner Asia that archaeobotanists could not trace one single path of distribution.

The wheat at Changguogou appears to be compact, and highly compact forms have been clearly identified in the area to the southeast, at Haimenkou in western Yunnan (1600–1400 BC).[48] Farther east, along the southern edge of the Himalaya in Nepal, highly compact wheats have also been recovered from Kohla (AD 500–1500).[49] The same morphotype appears to be present at

Burzahom, Gufkral, and Semthan in Kashmir (ca. 2800–2300 BC),[50] along with other southwest Asian domesticates. These crops spread into the region of Swat, Pakistan, likely directly from the Indus Valley region, on their way to Central Asia, by the early second millennium BC, as documented by finds at Ghalegay in Swat.[51]

HIGHLY COMPACT WHEAT

Compact Wheats on the Silk Road

While many subspecies and varieties of bread wheats are cultivated around the world today, two of particular interest are *T. aestivum* ssp. *compactum* and *T. aestivum* ssp. *sphaerococcum*. The grains of hexaploid wheats tend to be somewhat plumper than those of durum but have distinct rachises.[52] The grains of *compactum* and *sphaerococcum* are especially plump, the latter being nearly spherical. These ancient and lost varieties of wheat appear in assemblages from archaeological sites across Inner Asia and were clearly an important part of the economy along the ancient Silk Road. However, we know very little about where they originated, what their growing requirements were, or how they relate to certain varieties of modern wheat. It is possible that these ancient varieties were more drought tolerant than modern wheats, and they could have been among the keys to human adaptation to desert regions and possibly higher elevations. Early free-threshing wheats found in Central and East Asia tend to be small and hemispherical, a morphotype I refer to as highly compact wheat.[53] Similar wheats recovered from archaeological sites in the Old World have been referred to by a number of different taxonomic names, including *T. aestivum* ssp. *sphaerococcum*, *T. sphaerococcum*, *T. aestivum* ssp. *compactum*, *T. parvicoccum*, and *T. antiquorum*. The archaeobotanical grains from the burial cist at Begash and phase 1a wheat from Tasbas all express this morphology.[54] Highly compact wheat grains have been reported from archaeobotanical sites in both northern and southern Central Asia dating to between the late third and early first

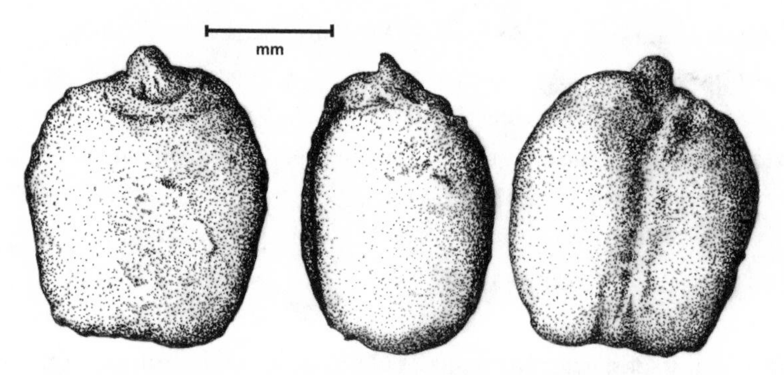

Figure 11. Three views of a highly compact wheat grain, dating to the middle of the second millennium BC, recovered from Tasbas, Kazakhstan.

millennium BC, and in southern Central Asia at Anau South and Gonur Depe by 2000 BC.[55] This highly compact free-threshing wheat persisted at Gonur Depe into the early first millennium BC.[56] Similar grains have also been recovered from Airgyrzhal-2 (ca. 2000–1500 BC) in the Naryn valley in central Kyrgyzstan.[57] Highly compact, round, free-threshing wheats were reported at Mehrgarh in the Indus Valley during excavations in the 1980s. Although these wheats were not directly dated, they may date to the middle of the fifth millennium BC, making them the oldest examples of their kind.[58] Other early examples of the highly compact morphotype have been reported from Harappan sites, ca. 2600–1300 BC,[59] and from the medieval Central Asian sites of Sapallitepe and Adytyntepe (ca. ninth century AD) in southern Central Asia.[60]

The mystery of what this highly compact wheat looked like, what its growing requirements were, and where it came from has received little attention until the past few years.[61] This may be due in part to the dearth of data from Central Asia in general, and in part to the difficulties of subspecies-level identification in the wheats. The lack of rachises in the finds makes it difficult to compare archaeological specimens with historically documented landrace varieties of highly compact morphotypes.[62] Furthermore, we have only cursory understandings of several other processes that could have led to changes in the morphology of the grains: for example, irrigating crops in arid regions affects the plumpness of the grains, especially when there are sharp transitions between dry and wet growing conditions.

In addition, studies have shown that heating and cooling grains can cause significant puffing and distortion, resulting in wheat assemblages that express a range of sizes and morphologies.[63] These studies are based on qualitative observations of length-to-width ratios, with no statistical analysis yet attempted. Scholars generally accept, however, that the short, round shape of the grains found at archaeological sites is too distinct to be the result of carbonization alone.[64]

A further complication is the wide variation in characteristics both within and between landrace varieties of any crop. Modern hybrid crops tend to be extremely uniform: some are cloned and thus express almost no genetic variation. By contrast, landrace varieties can express massive variation within a single field.

Discussing the Shortughai site in Afghanistan (2200–1500 BC), George Willcox notes that "given the array of varieties found in the region today the usual distinctions between forms break down because intermediates occur." However, he also showed that grains from some samples are generally more elongated, whereas grains from other samples are more spherical. He suggested that these differing morphologies constitute distinct genetic varieties and are not the result of environmental factors such as intensity of irrigation. Willcox also points out that the evidence from two of the samples "suggests that the crops were cultivated separately; perhaps one variety was suitable for dry-farming, the other better adapted to irrigated conditions."[65] Using length and width ratios, he identified two distinct varieties, compact and lax-eared. The lax-eared form is more elongated and narrower than the compact form. Archaeobotanists working in Europe tend to define the

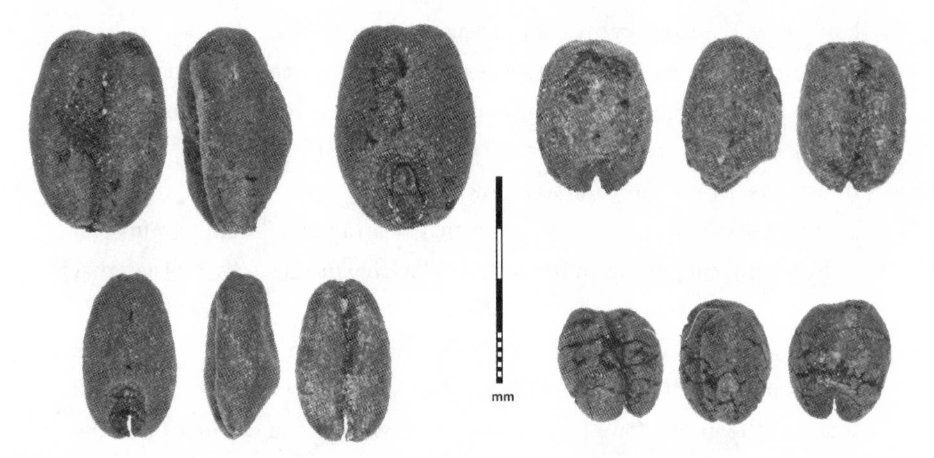

Figure 12. Four free-threshing wheat grains, all from FS7, a cache deposit at site 1211 in southern Turkmenistan, representing the range of variation present in one context.

compact wheats as those with a length-to-width ratio of 1.5:1 or lower, meaning that the distinction is not based on absolute length alone.[66] However, Willcox also showed that the range of variation in wheat morphology at Shortughai does not allow for clear divisions. This wide range of variation is characteristic of most landrace crops.

One trait that might help establish links between the historic landrace varieties and archaeological examples is the shallow ventral furrow in the grain. Ancient wheat grains that I have recovered from Tasbas and from the 1211 site in southern Turkmenistan (ca. 1500–1200 BC) have shallow furrows similar to herbarium and gene-bank specimens of dwarf wheat from India.[67] Based on the various classification attempts to date, a loosely identifiable variety of wheat appears to have been present across a large part of Asia during the late third and the second millennia BC.

The highly compact wheats are not a neatly defined category: they overlap greatly with other types of compact wheat. Buried ceramic vessels containing a mix of carbonized grains and legumes were found at the 1211 site.

One of the pots was filled mostly with carbonized wheat grains, although intrusive grains of barley and a few other crops were also present. This vessel contained nearly ten thousand wheat grains, providing a unique opportunity to study the diversity in a large population of wheat in Central Asia. However, the free-threshing wheat grains in the vessel range from nearly spherical and less than 2 millimeters in diameter (easily mistaken for millet) to elongated and up to 5 millimeters in length.[68] Figure 12 depicts the extremes of this variability. Hence the validity of any classification in Central Asia based on morphology requires further research.

Indian Dwarf and Japanese Semidwarfing Wheats

The significance of these highly compact archaeobotanical specimens of wheat, and the question of whether specimens from across Asia are genetically linked, remain fruitful topics of discussion. The debate becomes still more intriguing when scientists also consider herbarium specimens and extant landrace forms of highly compact wheats from northern India and Pakistan and the mountains of rural Japan.

Landrace forms of highly compact wheat were grown by a handful of isolated farming communities across Asia. Linking any of these populations of wheat on the basis of morphology alone is highly problematic, as a few scholars have noted.[69] Nonetheless, it is possible that the highly compact wheats found in Central Asia, at least, may have genetic linkages to forms that were once grown in Pakistan and northern India, notably the presumed ancestors of a landrace variety grown in that region before the Green Revolution, called Indian dwarf wheat (*T. aestivum* ssp. *sphaerococcum*).[70] According to historical accounts, this is a drought-tolerant variety, and this trait may have been the catalyst for its presence in Pakistan, Afghanistan, and northern India. Indian dwarf wheat is characterized by its short growth habit; however, it possesses a suite of distinctive traits, including strong, dense stalks and erect leaves, a condensed spike that expresses with short awns (the long, thick hairs at the end of each grain), glumes or protective shells around the grain, and a hemispherical grain. In addition, it has increased tillering (lateral branching) and a reduced rate of lodging (the tendency for the plant to bend over under the weight of its own ears). This distinctive suite of morphological traits in highly compact Indian dwarf wheats is referred to as the sphaerococcoid syndrome.[71]

Archaeological remains of highly compact free-threshing wheats, identified by archaeologists as Indian dwarf wheat, have been found at northwest Indian sites in Harappan and later levels.[72] If the identification and dating are correct, the oldest remains identified as Indian dwarf wheat are from the level III sediments at Mehrgarh (ca. 5500 BC).[73]

The dominant varieties of wheats in southern Central Asia varied over time. The earliest varieties, being glume wheats, were replaced by free-threshing wheat varieties that were easier to process into bread during the Chalcolithic, as evidenced by the assemblage at Monjukli Depe, an ancient village at the edge of the Kopet Dag Mountains and the Kara-Kum Desert in Turkmenistan.[74] This village, which was occupied almost six millennia ago, gives archaeologists a rare glimpse into early human life in southern Central Asia. Free-threshing wheat was the dominant form cultivated by Central Asian farmers from the fourth millennium BC on, and by the middle of the third millennium BC, free-threshing wheats seem to have become more compact than any of the earlier wheat forms.

Naomi Miller has pointed out that there is a chronological gap between the Neolithic site of Jeitun, which does not have highly compact wheat, and Chalcolithic Anau, which does.[75] She suggests that the highly compact wheat found at Anau, Djarkutan, and Gonur Depe (mostly in Bronze Age layers) could be related to Indian dwarf wheat, and that the time gap between these sites and Jeitun may indicate that the highly compact grains spread to southern Central Asia later, from the east (Mehrgarh or Pirak). The absence of such phenotypes at Sarazm, Tajikistan, in the fourth millennium BC lends support to this argument.[76]

While it is plausible that Indian dwarf wheat and the wheat remains recovered across Central Asia share an ancestor, the similarities in morphology are equally plausibly the result of parallel evolution. There were connections and exchange routes linking southern Central Asia and the Indus Valley during the third and second millennia BC.[77] However, tracing the spread of distinct varieties of wheat from the Indus Valley into Central Asia is complicated by finds of similar morphotypes in East Asia. Highly compact free-threshing wheat varieties have been discovered, in combination with barley, at archaeological sites in South Korea from ca. 1000 BC. Sites in Japan more than two thousand years old have also yielded highly compact wheat grains.[78] Some historical Korean landrace varieties, as well as many old Japanese varieties, also had highly compact grains.[79] Connections exist between the landrace varieties introduced to Korea and Japan as far back as the Mumun period (ca. 1500 BC) and small-grained remains found in China in the second millennium BC. These ancient peoples were moving from what is now mainland China through the islands off its coast, and they introduced new cultivated crops and domesticated animals to these islands. While it is tempting to connect these disparate data points across Asia and suggest links between these populations, the genetic data complicate the equation.

At first glance, the archaeological data seem to trace an obvious path for the spread of this compact wheat, from India, through Central Asia, and directly across northern China to Japan over the course of three millennia. However, before this reasoning can be considered conclusive, we need to understand the genetic basis for the morphological traits of the wheat grains that have been found and to determine whether their phenotypes represent a single broad gene pool or possible parallel (or convergent) evolution, likely through similar human-induced selective pressures. Some studies have attempted to understand the genetic basis of the collective suite of traits that make up the sphaerococcoid syndrome in wheat.[80]

Geneticists have mapped the *sphaerococcum* gene in hexaploid wheat; they specifically identified and described the wheat line with the genotype *ss*.[81] The mutation that caused this phenotype is likely the result of gene duplication during DNA recombination, which means that the plant accidentally duplicated a trait during hybridization, with a result that is visible in the entire plant.[82] This mutation likely arose relatively late during *T. aestivum* domestication (possibly calling into question the mid-sixth-millennium BC dates from Mehrgarh). Geneticists have suggested that early in the process of wheat domestication there would have been selection against drastic mutation syndromes, such as *sphaerococcum*, because of the undesired secondary traits associated with the mutation. However, these syndromes were selected for later on, possibly in the late fourth millennium BC.[83] It has subsequently been shown that the *ss* mutation controlling the length and width of the grain originated in India or Pakistan.[84]

The semidwarfing trait in most hexaploid wheats grown today is the result of selected alleles in multiple *Rht* genes. This means that there appear to be different genetic pathways toward the semidwarfing traits in wheat. There are 20 *Rht* loci and 25 alleles identified thus far, of which 11 occur naturally (14 alleles were obtained through induced laboratory mutations).[85] These genes are agriculturally important because they affect plant height, reduce lodging, and increase culm (stalk) strength, as well as increasing tillering; however, unlike *sphaerococcum*, they also increase seed yield. The breeding work directed by Norman Borlaug in the decades after World War II at the Centro Internationale de Mejoramiento de Maíz y Trigo (CIMMYT) in Mexico has become legendary, especially in India and China, where the effects of Green Revolution innovations to increase agricultural yields were most dramatic and most immediate. Borlaug built on the pioneering work of Nazareno Strampelli, an Italian biologist who isolated *Rht8* and haplotype I of the *Ppd-D1* from another Japanese landrace variety called Akakomugi. These two alleles, which marked for semidwarfing forms of wheat, became extremely important in his research into increasing plant yields. Strampelli first attempted to apply Mendelian genetics to wheat cultivation in 1900 by crossing two landrace varieties—Reiti and Noè—which expressed resistance to rust and lodging, respectively.[86] Ultimately, this line of research led to the breeding of the *Rht-B1* and *Rht-D1* alleles into wheat.[87] These two alleles were obtained from a Japanese landrace variety of wheat called Norin 10, and *Rht8* genes in Japanese landraces originated from a Korean landrace (*Anjeun baengyi mil*).[88] This genetic material is currently bred into over 90 percent of the wheat grown around the world.[89] The semidwarfing traits in cultivated wheats across China today have been linked to several different *Rht* loci, with *Rht-D1b* being distributed across the wheat-growing zones by means of government-run grain distribution programs. As much as 63.6 percent of the wheat in the northern summer-wheat-growing zone expressed the gene, compared with 43.5 percent nationwide.

While connections between ancient hemispherical, free-threshing wheat grains from Central and East Asia may seem obvious, there is no clear connection between the *ss* mutation and the *Rht* genes. Furthermore, a study conducted on the development of seedlings of Indian dwarf wheat concluded that its short growth habit is not the result of an *Rht* gene.[90] These findings suggest that morphotypical similarities between highly compact wheat varieties from Central and East Asia are the result of parallel domestication rather than common ancestry.

The Green Revolution's role in breeding *Rht* genes into wheat varieties drastically altered agriculture from the mid-twentieth century onward. Although the *Rht* genes used by both American and Italian breeders in the early twentieth century originated in Japan, it is unclear whether there are any linkages between the early Japanese semidwarfing wheats and Indian dwarf wheat.[91] Therefore, claims for widespread diffusion of this wheat morphotype across Asia are overhasty.[92] It is easy to make a simple, connect-the-dots argument based on archaeobotanical remains; however, additional genetic research is required before we can conclude that pre-Harappan farmers in India bred a phenotype that would form the basis of the Green Revolution.

The highly compact phenotype may not be the only set of genetic traits of wheat to have migrated through Central Asia. The original hybridization experiments conducted by Strampelli in 1900 used a little-known landrace variety called Reiti wheat, which contained the *Lr34* gene for rust resistance. This gene was subsequently bred into many cultivated lines around the world. But the gene has also been identified in historical landrace varieties from China and Central Asia; hence, a geneticist in Italy has theorized that wheat with the rust-resistance gene may have spread to Italy from the Black Sea region during the medieval period.[93] This theory is supported by the fact that several genes, such as *Ne1w* and *Bot(Tp4A)-B5c*, are found in both Reiti wheat and in Asian landrace varieties of wheat, but not in other European varieties. Thus it is possible that people who carried bread wheats from Asia all the way to Europe a thousand years ago are ultimately responsible for transferring rust resistance to most of the wheat crops in the world today. The Silk Road was the conduit for the transport of genetically improved lines of these crops, continually improving farming across the Old World.

ASIAN WHEATS: A SUMMARY

Free-threshing bread wheat is a major crop in China, second only to rice: the two are often grown in rotation on the same land, wheat in the winter and rice in the summer. While most people think of wheat as a European crop, Chinese cuisine would be hard to imagine without wheat dumplings and noodles. However, despite wheat's importance to the Chinese economy, debate continues over the origins, dates, and routes of its spread into China.

We know that wheat was domesticated in the Fertile Crescent roughly ten thousand years ago and that several forms of wheat made it as far east as Turkmenistan as early as 6000 BC. However, only free-threshing varieties of bread wheat seem to have reached China in antiquity.

In recent years, as a result of new archaeobotanical investigations in both Central Asia and China, it has been argued that wheat made its way into China around the late third millennium BC. The most plausible route is through the mountain passes that eventually became part of the great Silk Road.[94] Small-scale agropastoralists in eastern Kazakhstan may have passed wheat to closely related peoples in Xinjiang, from where it eventually passed through the Gansu Corridor into central China.

This early wheat was mostly highly compact and expressed a distinct morphotype. It may have originated in Pakistan or northern India and spread from there across Central Asia. While remains of similar compact wheats are found in archaeological sites in East Asia, their link to compact grains in Central Asia is unclear. Other forms of wheat, possibly with lax ears and a summer habit, may have been introduced into East Asia in the first or second millennium BC, along southerly routes of the Silk Road, such as along the southern rim of the Himalaya. The spread of wheat genes and crop varieties that reshaped Asian farming and culinary traditions was characteristic of the exchange routes of Central Asia, and it transformed economies across the Old World.

CHAPTER EIGHT

Legumes

With the exception of the soybean and the green bean or common bean, most of the familiar legumes in the Euro-American kitchen today originated in southwest Asia. However, the Euro-American focus on one small group of legumes, which includes peas, lentils, chickpeas, and fava beans, reflects the fact that agriculture originally spread into Europe from southwest Asia with crops that were domesticated in the Fertile Crescent. The range of plants in the Fabaceae family cultivated in China—beans, peas, vetches, and so forth—and the variation among many of the species is mind-boggling. Numerous species of legumes were domesticated in East Asia, and several more were introduced from elsewhere early in antiquity. In East Asia, where many people consume little meat and no dairy products, legumes are an important dietary source of both protein and calcium.

A number of these crops have regional or minor significance in China and were grown only in certain areas or on a limited scale. Examples include the adzuki bean or red bean (*Vigna angularis*) and velvet bean (*Mucuna pruriens*), both of which were likely domesticated in China and are grown only in a few regions today. The wild progenitor of the soybean is *Glycine soja*, which has been identified from a number of early predomestication farming sites in eastern China.

Many of the other species of legumes in China originated in far-off lands: the mung bean (*Vigna radiata*), sword bean (*Canavalia gladiate*), mat or moth bean (*Phaseolus aconitifolius*), and red or rice bean (*P. calcaratus*) are all believed to have initially been brought under cultivation somewhere in South Asia. The hyacinth or lablab bean (*Dilichos lablab,* also known as the dolichos or side bean) and black-eyed pea or cowpea (*V. unguiculata*) were likely first cultivated in North Africa. Shortly after European contact, the New World beans were also introduced to China. Another introduction was the yam bean (*Pachyrrizus erosus*), which originated in Central America but is sometimes called the Chinese potato today.

The diversity of landrace varieties of all these species in China is astonishing: for example, mung beans are usually green but can come in yellow or brown, and adzuki beans range from dark red to brown, yellow, and black. There are thousands of landrace varieties of soybeans, divided into three categories—yellow, green, and black—based on seed color.

Like most ingredients in the kitchens of East Asia, Fabaceae seeds and pods are prepared and cooked in an impressive variety of ways. Adzuki beans are used in several interesting Chinese dishes, including red bean paste, as a filling for steamed buns, and in a Cantonese sweet pudding. The unripe pod of the cowpea is often cooked as a green vegetable, similar to the American string bean; it is dry-fried in the Chinese dish of "yard-long" beans. However, despite the abundance of cultivated legumes in East Asia, and the diffusion of other East Asian crops along the Silk Road, there is very little evidence to suggest that these crops moved westward into Central Asia until after the colonial period.

SOUTHWEST ASIAN BEANS

The southwest Asian beans were part of a mixed farming system that developed around ten thousand years ago, which some archaeologists call the founder crop complex. This farming system relied on a combination of grain and legume crops (from the Poaceae and Fabaceae families, respectively). In the New World, the mixed cropping system included the "three sisters"—maize, beans, and squash. In East Asia, soybeans and rice often complement each other, and in the industrialized farming systems of the North American Midwest, soy and maize are paired.

There are two reasons that farmers throughout human history have paired grain crops with legumes. The first is that they are complementary in the human diet: the legumes provide protein, and the cereals provide storage carbohydrates that are easily broken down into sugars during digestion. The other reason is that they are complementary in the field as well: legumes have mycorrhizae, or nitrogen-fixing root bacteria (*rhizobia*), which put back into the soil the nutrients lost through the cultivation of cereals.

Despite the adaptive advantages of this mixed farming system, the first farmers to move into Central Asia for some reason left their legumes behind. Peas were introduced by the second millennium BC; lentils and chickpeas made it as far as Uzbekistan by the end of the first millennium AD, and lentils reached Kazakhstan shortly thereafter.[1] The first crops in Central Asia were primarily grain crops.[2] The archaeobotanical record is limited, but the absence of finds of legumes raises the question of why they do not seem to have spread with cereal crops through Eurasia, as they did through other parts of the Old World. The most obvious answer is that the proteins provided by legumes were not required in an economy that already incorporated dairy products and meat. The full answer may, however, be more complicated, especially when considering that legumes require far more water than grains.

A large number of the world's leguminous crops originated in southwest Asia. Many were part of the southwest Asian founder crop complex, such as peas (*Pisum sativum*), chickpeas (*Cicer arietinum*), grass peas (*Lathyrus sativus*), bitter vetch (*Vicia ervilia*), fava beans (*V. faba*), and lentils (*Lens culinaris*). Cowpeas are attested in Islamic contexts on the basis of archaeobotanical remains from northern Syria: these include one whole

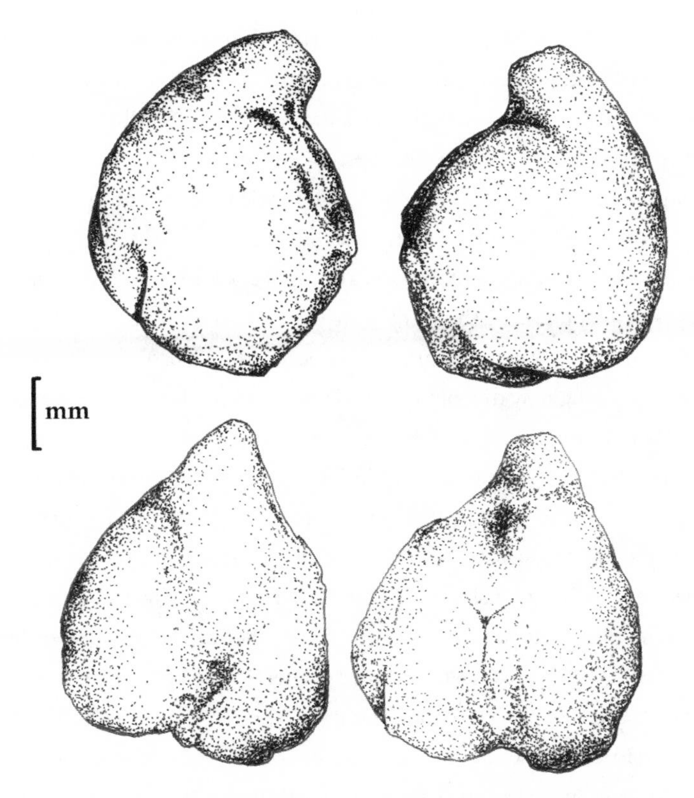

Figure 13. Four views of one chickpea recovered from the archaeological site of Tashbulak (AD 900–1200), in the foothills of Uzbekistan.

cowpea (two broken cotyledons) from a midden or trash deposit at Tell Guftan, dating to the late twelfth or early thirteenth century AD, and three fragments from a hearth at Tell Shheil I, dating to Abbasid layers in the ninth century.[3] While cowpeas clearly did not play a prominent role in cuisine, they were present in later Islamic Mesopotamia; however, as a frost-sensitive summer crop, they probably were not well suited to cultivation in Central Asia.

While all of these crops followed slightly different trajectories to domestication, most of them (except cowpeas) began the process around 9500 BC, along with wild *Triticum* and *Hordeum* grasses in the Fertile Crescent.[4] The domestication of legumes in southwest Asia may have been more rapid than that of grains.[5] One of the first traits of domestication in legumes must have been the breaking of seed dormancy. Most legumes have seeds that are programmed to remain dormant for a year or more before they germinate. Dormancy is an adaptation by the plant to ensure that some of the offspring survive years of drought or other poor growing conditions. In a bad year, all the plants that start growing may die, but the seeds that remain dormant in the soil will still be there to germinate the following year. Seed dormancy in wild legumes would have led to a low rate of germination and thus to a poor crop in any given season. Additionally, many wild legumes spread their seeds through a process called explosive dehiscence, in which the ripe seed pod pops open and shoots the seeds out. This obviously would have posed a problem for early farmers who wished to collect the seeds.[6] As cultivation practices selected against dormancy, the testa, or seed coat, became thinner. In addition, seed size and the number of seeds per pod increased; some varieties developed stiffer, more erect stems; and in a few cases toxins were bred out of the seeds. In some beans, such as the soybean, the toxins were never fully eliminated from the seeds, and fermentation or cooking is necessary to neutralize them.

On one of his legendary expeditions to trace the ancient routes of the Silk Road, Aurel Stein found a small cache of ancient grains and pulses that was dominated by a pulse that he referred to in his travel journal as *tarigh*. As Stein recalls in *Sand-Buried Ruins of Khotan*, he and his team dug these legumes out of the desert sands at the ancient city of Karadong on the Keriya River, along one route of the Silk Road halfway between Khotan and the Tarim Basin cities, in the middle of the modern region of Xinjiang in northwestern China. They cooked up the ancient legumes and found that the resulting paste was only good for sealing envelopes.[7]

One of the world's most popular legumes is the lentil; it is indispensable to most southwest Asian, Mediterranean, and northeast African cuisines. The wild progenitor for the modern lentil, *L. culinaris* ssp. *orientalis*, spans southwest and southern Central Asia, including the Fertile Crescent. The lentil is often grown in irrigated fields, like cereals, although it can also be grown on a smaller scale in gardens. Through domestication it branched into two morphological clades: small-seeded (*L. culinaris* ssp. *microsperma*) and large-seeded (*L. culinaris* ssp. *macrosperma*). *Microsperma* seeds are 3 to 6 millimeters in diameter, and *macrosperma* seeds 6 to 9 millimeters. M*acrosperma* lentils first appear in the archaeological record only about three millennia ago.[8]

The common pea, whose varieties include the snow pea, snap pea, garden pea, and the dried split pea used for soup, is another of the world's most popular legumes. The pea was domesticated from a thick-testa morphotype of the annual wild pea, *P. sativum* ssp. *fulvum*. Like the lentil, it can be traced back to the earliest Natufian plant cultivators in the Fertile Crescent well over ten millennia ago. There are now thousands of nondehiscent-podded landrace varieties of peas, ranging in height from one foot to several feet, with dense or weak stems, flowers ranging from blue and purple to white, and seeds ranging in size, smoothness, and color, from green to yellow, orange, and brown. This diversity of traits is what drew Gregor Mendel to the pea for his genetic studies, which ultimately transformed agriculture around the world. In early archaeobotanical remains of peas, the morphological traits for domestication are not as clear-cut as in other crops: the finds show a gradual increase in seed size and a smoothing of the seed-coat surface.[9] The last trait may be linked to a loss of dormancy in the crop.

The chickpea appears in the Euro-American diet most often in the form of hummus or sprinkled on a green salad, but it plays a much bigger culinary role in many other parts of the world, from India and Ethiopia to the Mediterranean. Domesticated chickpeas (*Cicer arietinum* ssp. *arietinum*) are self-pollinating and have pods with one or two seeds; the wild progenitor is most likely *C. arietinum* ssp. *reticulatum*. The chickpea, like the pea and lentil, was brought into cultivation across the Fertile Crescent around ten thousand years ago.

The fava or broad bean was likely also a member of the founder crops, although it is not found in abundance at many early sites.[10] Because its wild progenitor is unknown, it is hard to determine whether the archaeobotanical remains from southwest Asia represent domesticated specimens. A cache of 2,600 charred fava beans was recovered from Yiftah'el (8100–7700 BC) in Israel.[11] There are several varieties of fava bean, notably *Vicia faba* var. *minor*, with seeds measuring between 6 and 13 millimeters long, and *V. faba* var. *major*, with seeds between 15 and 20 millimeters long. The *minor* variety is more common in northern India, Pakistan, and Afghanistan; the large-seeded *major* forms may be a rather recent development.

Ancient fava beans were found at Mugh in the northern Zerafshan region, a fortified site at the base of the mountains that faces the Fergana lowlands. Excavators in the 1930s reported finding twenty-four specimens that they identified as fava beans.[12] The fortifications where these grains were recovered date to the seventh and eighth centuries AD.[13] However, because the original report claims that the beans have a peculiar angular shape and are almost equal in length and width, it is possible that they were actually grass peas (see below). The researchers also note that some of the specimens were black and some were a grayish color, and that they ranged from 8 to 9 millimeters in length, 7.5 to 8.5 millimeters in width, and 6 to 7 millimeters in thickness.[14]

Two other, largely forgotten leguminous crops originated in the Neolithic period in southwest Asia: the bitter vetch and the grass pea. The bitter vetch is high in alkaloids and is toxic if not processed by soaking, boiling, steaming, or leaching in pits of water or streams for long periods. The crop was used as animal fodder in Europe during the Roman period, and Pliny the Elder mentions its medicinal properties.[15] Like the bitter vetch, the grass pea is toxic if unprocessed: if consumed raw in large enough quantities, the neurotoxin it contains can cause lathyrism, a disease characterized by severe paralysis. This toxicity makes it all the more puzzling how early humans first learned to process the seeds. Like the bitter vetch, the grass pea was popular in Europe during the Roman period.

LEGUMES IN ANCIENT CENTRAL ASIA

Domesticated legumes were introduced into Central Asia from the Iranian Plateau around 2500 BC. Archaeobotanical assemblages from Namazgda culture sites in southern Central Asia included chickpeas (*Cicer* sp.), lentils (*Lens* sp.), and green peas.[16] The assemblages at Gonur Depe include several probable grass peas (*Lathyrus*).[17] Peas were found at Shortughai.[18] At site 1211 in Turkmenistan, dating to 1400–1200 BC, I sorted through a large cache deposit that included more than ten thousand peas, ranging in diameter from 3 to 7 millimeters and all with a smooth testa. The cache also included lentils.[19] As noted previously, I also recovered preserved peas dating to the middle of the second millennium BC from Tasbas, in the Dzungar Mountains of eastern Kazakhstan.[20] They are the only example of the southwest Asian legume crops to have been identified from northern Central Asian sites before the medieval period. Farther west, in the Khorezm Oasis, preserved grass peas have also been recovered at the site of Kara Tepe.[21] Peas are common finds in archaeological sites across South Asia, including India and Pakistan.[22]

The widest variety of legumes found at any archaeological site in Central Asia comes from Adji Ku, in the Murghab region of southern Turkmenistan, first excavated by the Joint Italian-Turkmen Archaeological Project.[23] This ancient city dates to the fourth millennium BC; however, the archaeobotanical remains that I analyzed in 2013 date to roughly 1900 BC. While most of the flotation samples from the site contained few preserved domesticated crop remains, one contained dense remains of wheat, barley, broomcorn millet, and legumes. Among the legumes were a single seed of bitter vetch, one fava bean, and one lentil, as well as two grass peas and several preserved peas. The lentil measured 3.0 millimeters in diameter; the embryo notch and the microbeak (the small protrusion from which the root shoot would have sported) were visible when the seed was held on its side. The preserved lentil is clearly lenticular in shape, meaning that it is semiflat and thicker in the middle than at the sides. While its small size does not allow us to rule out the possibility that it was a wild lentil, its association with domesticated grains and legumes suggests that it came from a cultivated crop. The two grass pea fragments recovered were from different peas. One whole cotyledon was measured, with a length (or width, as they are roughly square in shape) of 3.8 millimeters (figure 14e). The single specimen of bitter vetch measures 2.8 millimeters by 2.5 millimeters and has a distinct triangular shape in cross-section (figure 14a). The characteristic features are clearly visible on the archaeobotanical specimen. The single fava bean specimen measures 4.1 by 3.2 millimeters (figure 14c).[24]

In ancient Chinese sources, the common garden pea and the fava or broad bean share the vernacular name *hu tou*, the bean of the Hu people.[25] Hu is a familial name of people from Iran (ancient Persia or Iran) and southern Central Asia. The plant name thus indicates awareness of the bean's origins in Central Asia. The term *hu tou* was in use in the Chinese lexicon by the seventh century BC but may have had earlier roots, and peas were cultivated in the Tang and Sui Dynasties (AD 581–618).[26] Both fava beans and peas are widely grown in China today.

The fava or broad bean is also called the silkworm bean (*can tou*). According to one expert, "The broad bean may be a much later introduction, possibly coming eastward over the early Silk Road, that ancient caravan route linking China with the West."[27] Fava beans are not mentioned in any of the early Chinese sources and may not have been introduced to China until the Yuan period (1271–1368).[28] By the time of the Ming Dynasty (1368–1644), the fava bean was well known across China and widely cultivated, sometimes

Figure 14. Domesticated legumes from the Adji Kui site (FS 5.2) in southern Turkmenistan, dating to the early first millennium BC: (a) bitter vetch; (b) pea; (c) fava bean; (d) lentil; (e) grass pea. This site, from which I analyzed archaeobotanical remains in 2015, contains a more diverse variety of legumes than any other contemporaneous site in Central Asia.

on a large scale.[29] This hardy bean is grown instead of soybeans in mountainous and rainy regions of China.

Peas and lentils are often planted as winter crops in warmer parts of Eurasia, and in some parts of the world they are planted in autumn as a second crop. They are suited for winter crop rotations because of their frost tolerance and rapid growth. These traits may explain the presence of peas on the Himalayan Plateau and in the Dzungar Mountains in northwest China by the second millennium BC.[30] Evidence that peas and lentils were the first two cultivated legumes to spread north into Central Asia comes from

Djuvan-tobe and Karaspan-tobe, fortified ninth- and tenth-century villages on a tributary of the Syr Darya in southern Kazakhstan. Forty-one peas and four lentils found at Djuvan-tobe date to the ninth and tenth centuries. Eight specimens of carbonized peas recovered from Karaspan-tobe date to the fourth or fifth century AD.[31]

One more legume should be mentioned, even though it is essentially not known archaeobotanically in Central Asia, at least not in a domesticated state. Many scholars place the origins of alfalfa (*Medicago sativa*) in the Caspian steppe region. Wild relatives of the plant grow across the steppe and are found in Eastern Europe and Western Asia. One historian notes that the area of origin for the legume "included some of the steppe lands of Inner Asia from which pastoral nomads ventured forth repeatedly to pillage and to conquer. Indeed, the domestication of alfalfa in Western Asia was as a forage crop (first plant species to be grown as forage), and likely came about because of the growing importance of horses."[32]

Little is known about the origins and domestication of this crop. One possible archaeobotanical discovery of alfalfa comes from Kara Tepe in the Khorezm Oasis, south of the Aral Sea, dating to the fourth of fifth century AD.[33] It is mentioned in Hellenistic texts and, along with grapes, is one of the two crops that the mythical envoy of the Han Emperor was said to have brought back from Central Asia.[34] It was being cultivated in China as a vegetable for human consumption by the time of the Tang Dynasty.[35]

In the nineteenth century, Alphonse de Candolle studied the spread of the crop based on linguistic data; he noted that the English word *alfalfa* is derived from Arabic.[36] However, he references a thirteenth-century Arabic physician from Málaga in southern Spain who used a name derived from the Persian word for the crop, *isfist*. The Latin name is likely one reason it has long been thought of as a Central Asian crop: *medica* and the Greek word *medicai* are both references to the Medici of Media, in southern Central Asia. Interestingly, both Strabo and Pliny claim that the crop was brought to

Europe from Media during the Persian wars and that it was originally domesticated in Media as fodder for horses.[37] While there is no scientific evidence to support this idea, alfalfa was clearly known across Europe in the Classical period: it is mentioned by Varo, Columella, and Virgil.[38]

CONCLUSION

Peas had spread north into Central Asia by the second millennium BC, although legumes do not appear to have been a major crop in Central Asia until the historic period. Peas, lentils, and chickpeas are all important parts of the cuisines of Central Asia today. Chickpeas are added to pilaf, and lentils are often served as a paste, similarly to the way they appear in Turkic or Arabic cuisine. The peas found at Tasbas are possibly the best evidence for a second-millennium BC spread of agriculture along the mountain corridor. Peas are found in South and southern Central Asia but are absent from most of China, East Asia, and the rest of Central Asia.

The only other site where a pea has been identified is Changguogou in Tibet.[39] Tasbas and Changguogou lie at the ends of two different arms of the mountain corridor: Changguogou on the Himalayan Plateau and Tasbas in the Dzungar Mountains. A third arm of the corridor is the extension of the Pamir Mountains into the Kopet Dag Mountains and along the edge of the Iranian Plateau. Peas are found at the end of this arm, too, at site 1211, Ojakly. The fact that these crops have been found at three nearly contemporary archaeological sites that display little similarity in material culture and are separated by thousands of kilometers is the basis for the argument that I and my colleagues have made for the spread of crops along the mountain corridor during the late third and second millennium BC.[40]

CHAPTER NINE

Grapes and Apples

The emblematic cornucopia, or *cornu copiae*, the centerpiece of many dining-room tables, especially in European still-life paintings, is the divine "horn of plenty," a symbol of abundance, good tidings and the harvest, which originated in ancient Rome. The horn is associated with the earth goddesses Gaia and Terra; with a number of natural features, such as great rivers; and with Demeter, the goddess of the harvest. One account attributes its origin to the god Zeus, who as an infant was said to have been watched over by a goatlike nursemaid, Amalthea, while hiding in a cave on Crete from his ravenous father, Cronus. While suckling, the child inadvertently broke a horn off his surrogate mother. In another myth, Zeus's son Heracles broke the horn of Achelous, a personification of the largest river in Greece.

River deities are often depicted wielding a horn of plenty, overflowing with the fruits of autumn, in recognition of the prosperity that the rivers bring to the land and the fruits and grains that they nurture. On the Capitoline Hill in Rome, where the Capitoline Museum sits today, are two large white marble sculptures of muscular male figures, each bearing a cornucopia. One reclines on a sphinx and the other on a set of infant twins suckling from a she-wolf (figure 15). These figures are personifications of the Nile and Tiber Rivers. (Originally the Tiber figure was depicted reclining on a tiger and was intended to represent the Tigris River.) Aelius Aristides famously stated in the second century AD that to Rome "is brought from every land

Figure 15. Photos of the second-century AD sculptures of the personifications of the Nile (left) and Tiber (right), holding cornucopias, now residing on the Capitoline Hill in Rome. Photo by the author.

and sea all the crops of the seasons and the produce of each land, river, and lake, as well as the arts of the Greeks and barbarians, so that if someone would wish to view all these things, he must see them by traveling over the whole world or be in this city."[1] The fruits and grains watered by the mythical three thousand rivers of ancient Rome hailed from three continents, from the farthest corners of the known world. The cornucopia is often depicted in sculptures and murals holding peaches and apples, along with other fruits like grapes and pomegranates, which came to Rome by way of the Silk Road.

GRAPES

The most important fruit in the Roman cornucopia, inseparable from classical culture, was the grape, a symbol of the Greek god Dionysus (and the Roman god Bacchus). The simultaneously somniferous and terpsichorean effects of grape wine were esteemed across the ancient world. Ancient Chinese, Persian, or Arabic texts in praise of wine abound. Strabo pointed out the great importance of wine in Persia and further noted its production in Mariana, in southern Central Asia.[2] Herodotus too noted the significance of grape wine in Persia and among the Scythians.[3] Both authors commented on the Persian custom of discussing the weightiest matters of business and politics while both parties were drunk: if they both still agreed to the decisions the next morning, then they would act on them. This observation, by two ancient historians who are known to exaggerate, is supported by the writings of a Persian poet, Ferdousi, of the tenth century AD, in the *Shahnamah* (The epic of kings).

A recurring theme in both ancient Greek and Roman texts is the idea that Persian and other peoples of the ancient world were uncivilized because they drank their wine neat, whereas the Greeks and Romans mixed it with water to dilute the effects. They also note that the Persians engaged in Dionysian activities that apparently involved dressing like Scythians, dancing, and drinking abundantly.[4] Interestingly, Bacchanalian festivities are depicted on Sogdian tomb images from western China dating to the late first millennium AD, showing that as grapes moved along the Silk Road, so did the culture of wine fermentation and consumption.[5] It brought with it the gods of wine and the association with dance and lax social inhibitions. Yet the drink of Dionysus also has great significance in Christianity, Judaism, and Islam, and the drink of the lush is also the drink of the king.

Wines such as pinot noir, chardonnay, cabernet sauvignon, merlot, and even most of the brandies of Western Europe trace their pedigree to a wild vine in southern Europe and the western edge of southwest Asia. The affluent consumers of today pay exorbitant sums to purchase a wine with a certain terroir: a few years ago three bottles of Château Lafite's 1869 vintage went at auction for nearly a quarter of a million dollars each. Yet the grapes grown in Old World regions such as Bordeaux, Burgundy, Rioja, and Chianti are essentially the same as those grown in California, Oregon, Chile, Brazil, and New Zealand.

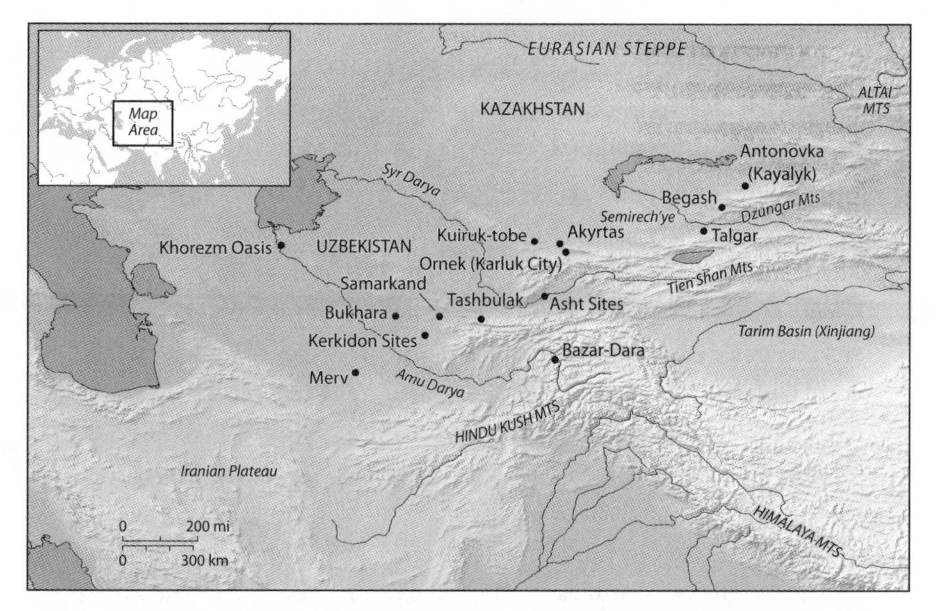

Map 4. Medieval archaeological sites, mostly villages or urban centers, that have provided archaeobotanical remains. Most of these sites contain remains of ancient fruits and nuts, illustrating the importance of the trade in fruit during the first millennium AD.

A Taste for Wine: Sogdian Merchants

The best-known cultural group of merchants on the Silk Road were the Sogdians, a people that originated in Central Asia but spread across China in the first millennium AD. In historical texts and artworks, they are often associated with grape wine, and viticulture became an important economic and cultural endeavor in eastern Central Asia and northwestern China at this time. The regions now comprising Xinjiang and northern Qinghai in western China were once referred to as the Western Regions. This area consisted of thirty-six kingdoms (or city-states) that later split up into more than fifty settled and fortified agricultural communities. Some of these outpost towns eventually became the oasis centers of the great Silk Road, allowing caravans to cross the Taklimakan Desert by following a stepping-stone path of oases through the dry sands. The grapes of Xinjiang were famous, and today many of these oases are still renowned for their grape harvests. Although other northern provinces of China, such as Ningxia, have become great wine producers for the global market, the history of China's wine industry is rooted in the oases of Xinjiang.

Traditional Uighur methods of wine production are still found in more remote regions of western China, especially around Kashgar. These wines are often referred to as *museles*, which means "triangle" in Arabic, and flavored with a variety of Central and East Asian fruits and spices. Many other key Silk Road fruits and grains are still cultivated in the oases of Xinjiang, including sorghum (which has largely replaced millet), barley, cotton, melons, jujubes, apricots, and pomegranates.

Turfan was a major oasis city along the ancient Silk Road and would have served as a link between dynastic China and Central Asia. The Turfan depression was an arable parcel of land that offered water and fresh foods to merchants and travelers crossing the desert. The tradition of wine production in Turfan, although now largely Europeanized, originated along the Silk Road routes in the middle of the first millennium BC. On four expeditions to study the Silk Road between 1900 and 1931, Aurel Stein and his teams excavated the remains of full vineyards, with preserved stumps of ancient grapevines, at the ancient village of Niya, just north of the modern town of Minfeng in the Tarim Basin.[6] Other artifacts from the site, including ancient Roman coins and textiles with classical motifs, secured the dates for Niya and the associated vineyards, which have subsequently been dated to the end of the first millennium BC. Some scholars have placed the earliest introduction of grapes to Xinjiang at roughly the third or fourth century BC. In support of this date, a tapestry recovered from tomb 01, room 01, at the Sampula cemetery bears a design that appears to represent clusters of grapes (figure 16). Radiocarbon dating of materials from this room provided a date close to

Figure 16. Sketch of a textile with a design that art historians have interpreted as grape clusters, unearthed from tomb 01, room 01, of the Sampula cemetery of Xinjiang, ca. first century BC. Adapted from Jiang et al, 2009.

AD 1. Although no grape seeds were found at Sampula, other important crops were identified in tombs from the same cemetery, including Job's tears, peaches, apricots, walnuts, Russian olives, broomcorn millet, and naked barley.[7]

Evidence for even earlier grape cultivation in Xinjiang has been recovered from a tomb at the nearby Yanghai cemetery in Turfan. The burials were spread along the edge of an oasis. A micromorphological study of a 116-centimeter-long branch from a grape vine recovered from a tomb at the cemetery shows that grapes were being cultivated in Xinjiang as far back as 390–210 BC. A number of other key Silk Road crops were also recovered from tombs in the Yanghai cemetery, including cannabis with elevated cannabinol levels; wild seeds from stone seed plants (*Lithospermum officinale*) used as decorative adornments; and capers (*Capparis spinosa*, presumably wild).[8] Nearly five hundred tombs were excavated at the site in the year 2003 alone.

The *Shiji* suggests that the legendary Han general Zhang Qian introduced grapes to China from Fergana in 128 BC.[9] When he returned from his arduous expedition to the west, he supposedly made reference to winemaking and claimed that wealthy vineyard owners in Central Asia could store up to ten thousand gallons (roughly 38,000L) of wine in their vats. This wine, he noted, could be collected and stored for several decades without going bad, thus providing an enduring form of wealth accumulation that other agricultural crops could not supply.

While it is clear that wine was in fact known in western China for centuries before Zhang Qian made his trek, he may have brought a specific variety of grapes back from the ancient kingdom of Dayuan, in the Fergana region of Uzbekistan, or he may have popularized the cultivation of grapes around the imperial center in Xi'an. The literary sources are unclear as to what kinds of alcohol were consumed in early East Asia, largely because of the varying translations of the word *jiu*. The *Shennong bencao jing* mentions grape wine production and is supposed to date to the first couple of centuries AD, but most historians believe this volume consists essentially of medieval accounts based on an earlier classic.[10] Likewise, a preserved account from the Wei emperor Cao Pi (Zihuan), from the Three Kingdoms period (AD 220–80), mentions the sweetness of grape wine in relation to grain alcohols.[11] Whether or not the failed mission to unite the pastoralist Yuezhi people of Central Asia with the Han Dynasty succeeded in bringing wine to Xi'an, the real wine producers and traders of ancient Asia, the immigrant Sogdians, took the stage a few centuries later.

Many written and artistic sources attest to the importance of wine in the cultures of Sogdiana (Central Asia) and the city-states of Xinjiang during the first millennium AD. Sogdian documents from the seventh- to eighth-century fortifications of Mugh in northern Tajikistan attest to large quantities of wine being produced for local consumption as well as sale.[12] The documents show that wine was often offered to guests and as a gift, payment, or tribute to political figures. Grape seeds were also found among a variety of other fruit remains in the citadel at Mugh.[13] The annals of the Xin Dynasty

(AD 9–23) note that grapes and wine production were abundant in Fergana and that the Sogdians enjoyed wine and dancing.[14] In a summary of early Chinese literary accounts, historians have noted many observations that wine was the main beverage in Fergana. The historical sources also discuss the establishment of a Sogdian colony in Lop Nor known as Grape City, which reportedly had a vineyard at its center.[15] A specific account attributed to Emperor Yuan of the Liang Dynasty (r. AD 552–55) mentions that the Yuezhi were winemakers, although this may simply be drawing on descriptions in the *Shiji*.[16] The *Hou Hanshu* (Book of the later Han), compiled by Fan Ye (AD 398–446), noted that grapes were grown in Hami, a town in Xinjiang, along with rice, both millets, wheat, legumes, mulberry, and hemp. Further along, in a section titled "Kingdom of Liyi," which most scholars believe to be Sogdiana, he notes that horses, cattle, sheep, grapes, all sorts of fruit, and wine were produced.[17]

Wine continued to be an exotic commodity in central China until the Tang Dynasty. In the fifth century AD grapes and wine were still imported from western China.[18] Even before then, however, wine was important in politics and among the elites. Stories dating back to the middle of the first millennium BC recount government officials being bribed with rice beer or wine, the presentation of wine as an offering, and its use as a means of poisoning a political opponent or simply to intoxicate a rival during negotiations. In one of these accounts, from the *Hou Hanshu*, Meng Tuo at Fufeng bribes a government official with one *hu* (about twenty liters) of wine to obtain an honored political position in what is now the region of Gansu and Ningxia. Tales of assassination attempts through poisoning were commonplace.[19] A number of kings or important political figures were said to spend most of their time intoxicated.

One of the clues that wine was regularly being imported from Central Asia to the dynastic centers during this period is the large number of highly ornate gold and silver drinking bowls that have been recovered in China dating to between the third and the mid-eighth century AD. These goblets and bowls are decorated with Central Asian motifs and, in some cases, Classical imagery, such as acanthus leaves.[20] These wine bowls became prevalent during the Tang Dynasty, after military advances into Central Asia put much of the Silk Road under Chinese control. Some of the best examples of Sogdian or Central Asian-style drinking bowls come from the Hejia village hoard, from just outside the former Chinese capital of Chang'an. Similar vessels are depicted in tombs and on Buddhist wall art from across northwestern China in association with Hellenistic images (figure 17). Aurel Stein, digging at the Buddhist ruins of Miran in February 1900, describes friezes on the wall in the vaulted chapel enclosing a stupa, dating to around the fourth century AD. These images contained Greco-Roman putti, festival scenes, and youths with goblets of wine.[21]

During the Tang Dynasty, as wine became popular across the Chinese Empire, it became the drink of the artisan and poet. Countless examples of early Chinese poetry refer to wine (or alcohol generally) and the loosening of inhibitions that results from consumption. The legendary poet Li Bai (AD 701–62) wrote at great length about the virtues of wine, especially its ability to ease sorrow; one of his most famous poems is titled "Waking from Drunkenness on a Spring Day." Li Bai was one of the Tang Dynasty poets from Chang'an known as the Eight Immortals of the Wine Cup because of their drinking habits. The rapid increase in popularity of grapes during this period may have been linked in part to the empire's prosperity and a rising demand for exotic goods.

Another contributing event may have been the Tang conquest of the Gaochang oasis city-state near Turfan, in AD 641, which solidified the imperial control of exchange on the Silk Road. This may have led to the spread of a legendary landrace variety of grapes from Central Asia into China: "mare's nipples," cultivated widely in Taiyuan in Shanxi and celebrated in Tang poetry. An account from AD 647 describes this grape variety as having clusters of purple fruit two feet long.[22] The *Tang Shu* (Book of Tang, chapter 200)

Figure 17. A carved stone panel from a Tianshui funerary couch, from a Sogdian burial in Gansu dating to the sixth or seventh century AD. Art historians interpret the image as a Dionysian party, with wine pouring out of fountains and people drinking from ewers. The panels are currently displayed at the Municipal Museum Tianshu. Photo by Patrick Wertmann.

claims that the mare's nipples grape was given directly to Emperor Tai Zong (AD 598–649), along with instructions for making wine; the emperor then passed the grapes and knowledge to his people. Xinjiang and Central Asia were still credited with producing the best wine, and several odes are preserved lauding the quality of wine from Liangzhou in Gansu.[23]

The growing cosmopolitanism of the Tang Dynasty was marked by wine-infused parties accompanied by music and featuring dancers from Central Asia, who were considered exotic and greatly praised in the dynastic center. In the Tang capital there were wine shops, notably in the walled-in western market, one kilometer square, in the Liquan ward of the city.[24] The market's Persian sector boasted bars and wine shops owned by Sogdians and featuring Central Asian performers. These shops were epicenters for cultural exchange and trade negotiations.

The Tang solidified control over the western regions of the Silk Road by increasing military power in the area of modern-day Xinjiang. As exotic goods poured into the empire, it was no longer only the elite who enjoyed wine from goblets: even soldiers and commoners drank Central Asian wine. The eighth-century poet Wang Han imagined a soldier's experience:

> Holding a glowing goblet filled with grape wine,
> Following the melody of a lute, I am about to drink,
> The neighing horse urges me to ride him.
> Do not laugh if you see me lying drunk on the battlefield,
> Few soldiers ever come back from western expeditions anyway.[25]

Among the funerary artifacts from several Sogdian tombs in China, scholars have noted funerary platforms, or couches, with painted or carved iconography. One of these, from Tianshui in Gansu, dating to the sixth or seventh century AD, features a painted and engraved panel depicting a scene of a Dionysian party as well as wine production, an image taken from late Roman mythology (figures 18 and 19). Art historians have linked classical imagery in China with other Silk Road images, such as the "Zoroastrian Praises from Ancheng," an ancient preserved document from Dunhuang, Gansu.[26] The Tianshui panel features a vessel with wine pouring into it from two gargoyle-like fountainheads next to a Zoroastrian fire altar in a temple. A very similar scene, also mingling Classical and Zoroastrian traditions, is depicted in a mural from Panjikent. Many other similar images and literary references from China in the late first millennium AD represent Central Asian musicians and whirling dancers in association with the consumption of grape wine from goblets and ewers.

Wine and Vinegar: A History

Thirty-six species in the genus *Vitis* grow in China, most of which are restricted to southern China.[27] Of these, the only one of economic significance is the Amur grape (*V. amurensis*). Cultivated in Russia and northern China, it is highly frost tolerant and recently has been crossed with European grapes to produce hardier lines. This grape was probably brought under cultivation in northeastern China only within the past century. Archaeobotanical evidence from regions of southern China clearly shows that people were collecting at least some of these wild grape species for millennia before the introduction of European grapes.[28]

The European table grape, the source of all European wines, descended from *Vitis vinifera* ssp. *sylvestris* in southern Europe and southwest Asia. Modern European table grapes tend to be seedless and asexually propagated, a fact that complicates genetic and archaeobotanical studies. The domesticated table grape is also hermaphroditic, while its wild ancestor is dioecious (having male and female flowers on separate plants); it comes in a wide range of shapes, colors, and levels of sweetness.

Because of its cultural and economic importance, the European grape has received considerable attention from archaeologists, historians, and geneticists. A large-scale genetic study in 2010 of herbarium and gene-bank

Figure 18. A carved stone panel of a Sogdian funerary couch from Anyang, likely created during the Northern Qi Dynasty (AD 550–77). The image shows a celebration taking place under a grape trellis, with clusters of grapes hanging down and a man at the center dressed in Central Asian attire, drinking from an ornately decorated drinking horn. Photograph © 2018 Museum of Fine Arts, Boston.

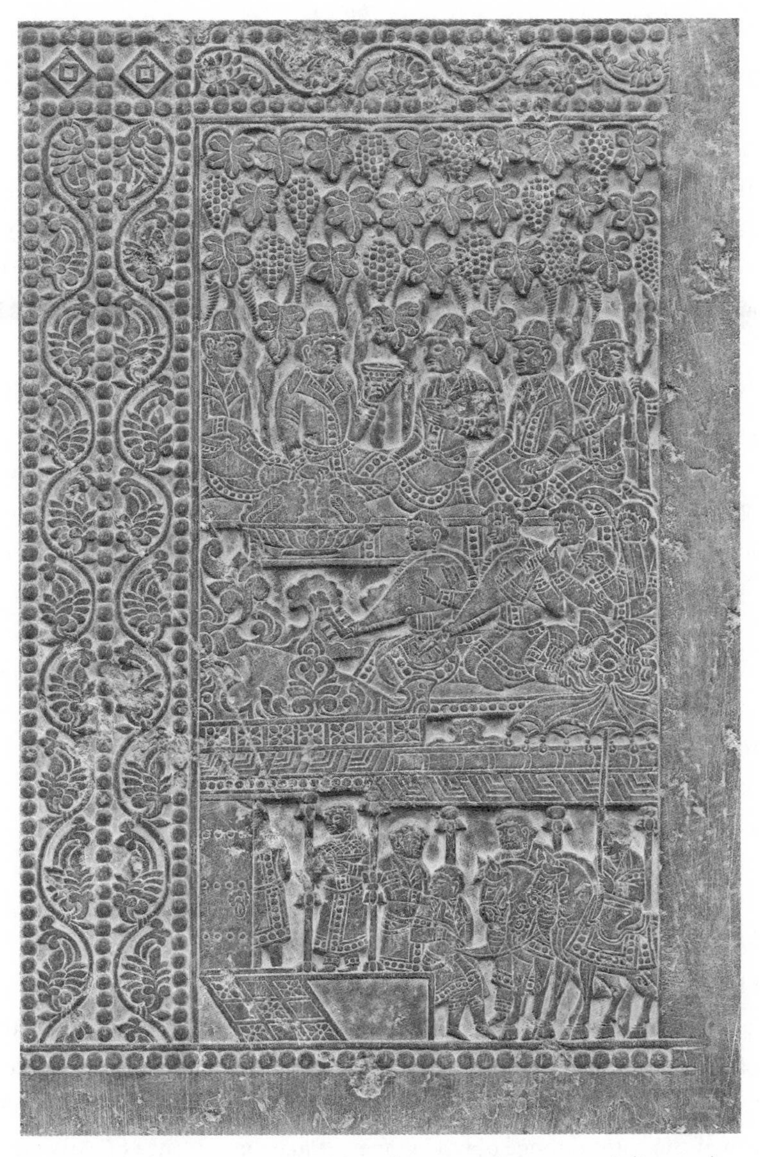

Figure 19. The opposing stone panel from the Anyang funerary couch (figure 18). The top row of figures represents a drinking celebration under a grape trellis; below, Sogdian performers play typical Central Asian musical instruments. Wine would have been an integral part of these social and political events. Photograph © 2018 Museum of Fine Arts, Boston.

specimens of grapes looked for genome-wide patterns and genetic variation within the cultivated and wild populations. This study supported the idea that grapes were first cultivated in southwest Asia, but it also identified a lot of genetic crossing with wild lines as the crop moved into Europe. The geneticists also argued for a weak domestication bottleneck, meaning that there is a close relationship between wild and cultivated populations, and millennia of asexual cloning of grapes that maintained distinct cultivars. In the long term, this practice was a double-edged sword, fixing desirable traits in favored cultivars but also stagnating the plant's genetics and reducing diversity, leading to the increased vulnerability to pests and diseases that has periodically decimated the modern wine industry.[29]

In an influential paper aptly titled "Sweeter than Wine," Naomi Miller argues that the grape was originally cultivated not for the making of the intoxicating beverage but for its sweetness, equally precious in a world without cane sugar.[30] Residues of tartaric acid, indicating the presence of wine, have shown that wine was being stored in ceramic vats in southwest Asia by the middle of the sixth millennium BC, although Miller points out that this wine was likely produced from wild grapes, and viticulture was probably not prevalent for another three millennia.[31] The tartaric acid residue on another vessel from the archaeological site of Hajji Firuz, in the Lake Urmia basin of northwestern Iran, provides some of the earliest evidence for wine production (although contemporaneous evidence has been recovered from as far away as Georgia).[32] This vessel once held nine liters of liquid, and another vessel from a nearby site held as much as fifty liters. On the basis of the volume of these vessels, scholars have argued that they held wine and not vinegar, because every household could use fifty liters of wine but not fifty liters of vinegar. Fourth-millennium BC tartaric acid residue was also recovered from ceramics at Godin Tepe in Iran. Because this area is beyond the natural range of wild grapes, it may be evidence of early grape cultivation.[33]

Archaeobotanical remains of grape seeds from southwest Asia and the Iranian Plateau date all the way back to the Pleistocene. Grapes have attracted human attention since the first hominids moved into the Mediterranean, and they were undoubtedly collected from the wild liana long before they were domesticated. Determining the date of the earliest cultivation of grapes is complicated, however, because the morphological indicators of domestication—such as increased sweetness, the amount of flesh on the fruit, hermaphroditism, and the number of fruits in a cluster—cannot be detected from archaeobotanical evidence. Grape seeds are often well preserved, but the seeds of wild and early cultivated grapes are morphologically indistinguishable.

The oldest macrobotanical evidence for grapes in Central Eurasia comes from excavations of Namazga V and VI Culture levels at Anau South in Turkmenistan (ca. 2500 BC).[34] Other Namazga V sites in southern Central Asia (ca. 2000 BC), including Gonur Depe and Djarkutan, have yielded grape pips, showing that viticulture was established in southern Central Asia by the late third millennium BC.[35] At Mehrgarh in Pakistan, Harappan viticulture is also attested by 2000 BC, based primarily on the presence of grape wood.[36] Archaeobotanists have also identified grape wood dating to 1700–1000 BC at Burzahom in Kashmir.[37] Fragments of grape pips were found at the sixth- to fourth-century BC Achaemenid city of Kyzyltepa, in the Surkhandarya region of southern Uzbekistan.[38] Numerous grape pips were also recovered from mudbrick fragments at the fifth- to seventh-century AD sites of Kuyuk Tepe, Munchak Tepe, settlement 5a, and Tudai Kalon, all in the Osh region of Kyrgyzstan.[39] Grape pips, pedicels (the stalks that support the fruit), and even a whole carbonized grape were found preserved in a flotation sample from a midden deposit in the center of the Silk Road town of Tashbulak in Uzbekistan dating to the late first millennium AD (figures 20 and 21).[40]

Farther north in Central Asia, numerous grape pips from different excavation seasons at Tuzusai, Kazakhstan, indicate the existence of viticulture in the northern routes of the Silk Road by the fourth century BC. The

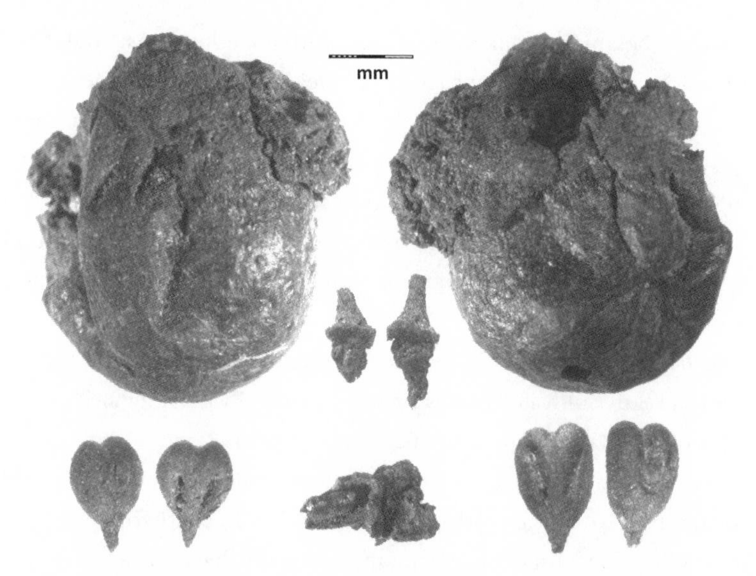

Figure 20. Front and back views of a full preserved grape, with two pedicels and a broken seed with grape flesh still attached (bottom center), and ventral and dorsal views of two grape pips (bottom left and right), all from Tashbulak, Uzbekistan (AD 900–1200).

abundance of pips suggests that grapes were cultivated locally.[41] In addition, there are vineyards on the Talgar alluvial fan, near the Tuzusai site, today. If viticulture was being practiced in northern Central Asia during the transition from the Bronze to the Iron Age, it not only means that investment in plant cultivation was significant, it also suggests a concept of land tenure completely different from what most scholars have assumed existed in the region. Grapes are secondary crops, which are usually brought into an economic system only after primary staple crops are well established.[42]

Soviet scholars have identified a handful of what they have interpreted as wineries dating to the first millennium AD. In a few cases these wineries are also associated with finds of preserved grape pips. One of the best examples is at settlement 5a in the cluster of Kerkidon settlements in the Osh

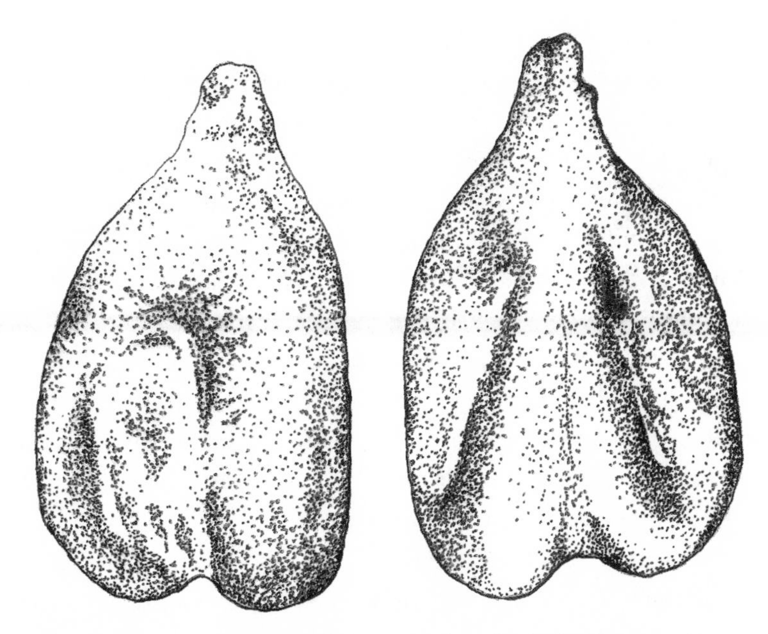

Figure 21. Scientific illustration of two views of a carbonized and preserved grape pip from Tashbulak, Uzbekistan (AD 900–1200), FS25.

region of Kyrgyzstan. The archaeologist working on the project reconstructed the winery based on ethnohistoric descriptions of wineries in Samarkand in the late 1800s.[43] He claimed that willow or camelthorn (*Alhagi* sp.) branches were piled on top of two parallel support beams, and the grapes were laid on top of them for pressing. The grapes were pressed by foot, and the juices flowed along drains into large sediment vats. The Kerkidon winery had two large sediment vats (holding about four hundred liters each) buried in the ground at the end of two drain channels made from fired bricks. The oldest of these supposed wineries, in southern Central Asia, dates to the Kushan Empire (second century BC–third century AD).

Other examples of wineries across Central Asia date to the medieval period: these include one from outside the Silk Road city of Penjikent in the

Zerafshan River Valley in Tajikistan, one at the Lugovoye B settlement in Semirech'ye, and one outside Saryg in the Chuy Valley of the Tien Shan Mountains of Kazakhstan.[44] Soviet excavations in the residential areas of ancient Penjikent, dating to the seventh and eighth centuries AD, identified a large number of wineries with wine presses and large ceramic holding tanks.[45] In addition, massive ceramic vessels (280–300 liters) intentionally sunk into the ground are found at urban sites across southern Turkmenistan; some of the best examples come from Nisa near modern-day Ashgabat, but they are also common at Ulug Depe.[46] These finds are consistent with the winemaking process described by Omar Khayyam (d. 1131), one of the great thinkers of medieval Persia. He refers to pressing grapes and letting the juice flow into holding vats, noting that the wine will bubble like water in a boiler, but without fire—the fermentation process.[47]

Many of the vessels are found inside house structures, and while it is possible that some of them were used for water storage, there are several with ostraca (inscribed fragments of ceramic pots) containing writing that suggests they may have been used for wine storage. The ostraca at Nisa date from the mid-third century BC to the mid-third century AD and contain a number of written references to wine, its quality, its age, and whether it has turned to vinegar.[48] A few even refer to a white or rose-colored wine. (A few ostraca also mention raisins, flour, oil, flax seed, sesame seed, wheat, and barley).[49]

Other inscriptions referring to wine drinking appear on the rims of Sogdian wine vessels called *khums* that have been recovered from across Semirech'ye; many of these date to around the eighth or ninth century. One of these inscriptions, from Krasnaya Rechka, excavated in 1988, states, "He who has counted no loss will also not see his wealth, so, O Man, drink [wine]!"[50] Another, recovered in 1941 about twenty kilometers west of Krasnaya Rechka, close to the site of Novopokrovka, states, "May this wine be drunk [at] a happy moment."[51] Other Sogdian inscriptions in Central Asia, from sites such as Penjikent and Mugh, also mention wine or the sale of grain or bread.

Figure 22. A young girl selling grapes in the Bukhara bazaar in 2018; she is displaying a small grape variety from Samarkand. Hundreds of landrace varieties of grapes are maintained in Central Asia today. Every produce market offers numerous varieties, each with its own unique flavor and texture.

The ancient city of Nisa, founded by Arsaces I (r. 250–211 BC), was for a while the central political city of the Parthians. The site is located about eighteen kilometers from Ashgabat, on the boundary between southern Turkmenistan and Iran. It would have served as a significant stop along the Silk Road during the Roman period, and its lavish art, architecture, and material remains reflect a strong Hellenistic influence and the lasting legacy of Alexander the Great. Massive Soviet excavations were conducted at the site from the 1950s to the 1970s, and an astonishing array of elite items has been recovered.[52] They include a large number of rytons (drinking horns) made from ivory, with detailed gold and silver decorations along the rim and at their base (see figure 18). Their craftsmanship illustrates the ritual nature of wine drinking in elite circles. Their adornment with images of Olympian gods, mythical animals, and other Greek figures shows the persistence of a strong Hellenistic influence in Central Asia and the close cultural ties between Bacchanalian revelry and the Central Asian tradition of wine drinking, which continued until well after the Arab conquests.[53]

Other decorative evidence of viticulture along the Silk Road includes a motif of clusters of grapes stamped into the plaster of a door frame in the citadel at Penjikent.[54] During the medieval period this would have been a highway of traffic, especially during the Qarakhanid reign, when the empire was divided into two halves, with capitals in Samarkand and in Kashgar. Among the exotic exchange goods excavated from Penjikent are Arabic documents attesting to mercantile activities, and coins from China and from across Central Asia.[55] An identical ornamental grapevine doorframe recovered from medieval Bukhara is on display in the city's Ark Fortress museum.

Grapes remained an integral part of southwest Asian cuisine even after the Arab conquests and the Muslim proscription against alcohol, and many of the early geographers discuss grape cultivation. The tenth-century geographer Ibn Hawaqal notes the prevalence of vineyards in the upper Euphrates region, as does al-Muqaddasi.[56] Ibn al-Awam discusses the proper way to cultivate grapes. Many of these accounts note that vineyards were irrigated and that the quality of the fruit was improved by cutting off irrigation shortly before harvesting. From southwest Asia up to Xinjiang, vines were cultivated on high trellises, which not only provided shade for people in the hot summer months but also shaded the delicate fruits. Banquets were often held under the shade of a grape trellis. In medieval Asia, grapes were processed into raisins or syrups called dibs; they were also consumed fresh and, contrary to later Islamic custom, used to produce wine.[57]

Viticulture was established in southern Central Asia by the late third millennium BC. It took roughly two thousand years longer to become established in northern Central Asia. However, grapes eventually became the most highly lauded fruit of the Silk Road, cultivated in the desert oases and on the low foothills of Inner Asia. Grape wine became one of the most important commodities of the ancient world, allowing people to accumulate an agricultural surplus that could be stored for decades and to display their wealth. Thanks to the Silk Road merchants who carried skin sacks of wine, vats of dried raisins, and the genetic material for the most prized grape cultivars along the trade routes through Central Asia, wine is now an essential component of the culinary traditions of Europe and Asia.

APPLES

The Story of the Forbidden Fruit

> Of man's first disobedience, and the fruit
> Of that forbidden tree, whose mortal taste
> Brought death into the world, and all our woe,
> With loss of Eden, till one greater Man
> Restore us, and regain the blissful seat,
> Sing, Heavenly muse . . .
>
> John Milton, *Paradise Lost*

Neither John Milton nor the author of Genesis identified the forbidden fruit in the Garden of Eden as an apple; however, European mythical tradition abounds with references to the power and menacing associations of the apple. Later in *Paradise Lost*, Milton does refer to the apple, with a play on words linking the Latin genus name, *Malus*, to malicious intent. To Paris of Troy, the apple is a sexual token. The jealous stepmother casts a somniferous curse upon Snow White with a poisoned apple.

More recently, the apple was rehabilitated in popular Euro-American culture. Its propagation is part of the mythology of the conquest of the American West, thanks in part to Johnny Appleseed (Chapman).[58] Fermented or hard apple cider, long popular in northern Europe and the United Kingdom, became an important part of the early colonial economy in America, and every European settler planted one or a few trees for cider production. The apple, and hard cider generally, lost popularity during Prohibition (1920–33) in the United States. Eventually the shiny red apple was cleaned up in American culture, losing the alcoholic legacy and gaining a wholesome reputation. The apple became the stereotypical gift to teachers, and its health-giving properties kept the doctor away. In one of several popular scientific books that have explored the history of the apple, Thor Hanson points out that the fruit went from "a single species domesticated in the mountains of Kazakhstan to thousands of varieties—people grow them on every continent outside Antarctica."[59]

In his well-known essay "Wild Apples," Henry David Thoreau observes, "It is remarkable how closely the history of the apple tree is connected with that of man."[60] Indeed, the relationship between apples and people is close and complex, spanning at least five millennia. The story of the apple begins along the Silk Road. American apple pie, apple juice, apple cider, apple cobbler, apple fritters, apple butter, and apple preserves all trace their origins back to a few river valleys in the Tien Shan Mountains. In recent years genetic studies have resolved much of the debate over these origins. Nevertheless, the ancestry of the apple is highly complex. Cloning, inbreeding, and reproduction between species have created a genealogy that looks more like a spider's web than a family tree. To growers, the beauty of the apple lies not in its rosy (or green or yellow) skin but in its genetic variability and plasticity, its ability to cross with other species of *Malus* and other distant lines of *M. domestica/pumila*, and the ease with which it can be grafted onto different rootstocks and cloned.

While the apple in your pie is likely from one of just a few dozen commercially popular cultivars, grafted onto one of about ten widely used rootstocks, the genetic diversity of the genus *Malus*, both wild and cultivated, was and still is immense. Our modern large-fruited, sweet, and abundantly producing varieties can trace their origins to the Silk Road. The true ancestor of the modern apple is *M. sieversii*, the species romantically described by Michael Pollan in *The Botany of Desire*.[61] Remnant populations of wild apple trees survive in southeastern Kazakhstan today, producing fruits with considerable phenotypical variation: they are roughly eight centimeters (just over three inches) in diameter and range from green to red in color and sweet to tart in taste. The geneticists who worked on a large-scale study to map the genome of the apple and to understand its domestication state that people carried the first domesticated apples "westwards along the great trade routes known as the Silk Route, where they came into contact with other wild apples, such as *Malus baccata* (L.) Borkh. in Siberia, *Malus orientalis* Uglitz. in the Caucasus, and *Malus sylvestris* Mill. in Europe."[62] Hybridization among these distantly related genetic lines, genetic bottlenecking (the isolation of a small group of trees from the larger population) as populations were transported on the Silk Road and eventually around the globe, and human selection for desired traits created the Baldwin, Braeburn, Cameo, Cortland, Crispin, Empire, Fortune, Fuji, Gala, Golden Delicious, Golden Supreme, Honeycrisp, Jersey Mac, Jonathan, Jonagold, Jonamac, Macoun, McIntosh, Red Delicious, Rome, and Ruby Lady, to name but a few of the thousands of extant landrace varieties. Grafting created stronger trees that could carry larger loads of fruit, and cloning, by taking cuttings, allowed for the fixation of varieties and preservation of desired traits.

The genetic contributions from the wild apple relatives, which we would call crabapples, were not trivial. *M. sylvestris* made an especially important

contribution to modern varieties as the tree moved westward along the Silk Road.[63] Self-incompatibility in apples (the inability of a tree to self-pollinate, meaning that at least two trees are needed to produce fruit) likely enhanced the gene flow between wild and cultivated species as the trees moved across Europe and Asia. Ultimately the covenant between humans and crabapples (entered into some five thousand years ago between prehistoric Central Asians and a wild population of *M. sieversii*) led to one of the most profitable and widespread fruiting tree species on the planet today.

Parsing out the details of this coevolutionary process has proved difficult, largely because of a dearth of evidence. Only a few archaeobotanical remains of ancient apples have been found along the Silk Road, although genetic, linguistic, and historical evidence illustrates that they were abundant. Furthermore, little work has been done on ancient carbonized seeds to differentiate between wild apple relatives and cultivated variants of apples. The story is further complicated by the fact that two other close relatives of the apple were also domesticated in Central Eurasia and spread along the Silk Road: the quince (*Cydonia oblonga*) and certain varieties of pear (*Pyrus* spp.), whose seeds are morphologically very similar to those of the apple. Very little is known about their narratives, although we know that pears were moving along the Silk Road by the Tang period. Whole, desiccated fruits were recovered from the Astana cemetery in the Turfan region of Xinjiang.[64] Several species in the genus *Pyrus* were cultivated in China, and by the Tang Dynasty pears were a common fruit. The pear was also common in the Mediterranean in the Classical period: it is depicted on murals at Pompeii as well as in the writings of Homer, Theophrastus, and Dioscorides under various derivatives of the word *apios*, and by Pliny under the Latin name of *pyrus*.[65] It is also possible that the medlar (*Mespilus germanica*), another member of the Rosaceae family, had a wider ancient population, possibly stretching to West Asia, but almost nothing is known of its origins, and today it is wild only in Germany.

Archaeobotanical remains of crabapples have been recovered from early archaeological sites across Europe. These include whole fruits as well as fruits that were cut in half and strung up to dry.[66] It is not possible to ascribe these remains to distinct species, although they closely resemble genuinely wild species that grow across Europe today. Some wild forms of *M. sylvestris* have large fruits, but heavy gene flow between domesticated and wild forms complicates our understanding of these supposed wild populations. Moreover, these discoveries do not necessarily tie into the narrative of our modern apples, because many preserved crabapple fruits and seeds are from truly wild species and in some cases predate the spread of *M. domestica/pumila* into Europe. Ancient Europeans were eating crabapples for many millennia before true apples were known. The presence of crabapples in the human diet may have facilitated the hybridization between wild *M. sylvestris* and domesticated lines.

One significant archaeobotanical discovery is a string of halved, desiccated apples found on plates in Queen Pu-abi's grave at a cemetery near the ancient city of Ur at the mouth of the Euphrates River (near present-day Basra), dating to about 2200–2100 BC. These fruits are likely wild crabapples, though they may represent undomesticated cultivated apples.[67] The dried fruits were roughly eleven to eighteen millimeters in diameter.[68] Viticulture and even arboriculture were already known to the Sumerian people of the capital city of Ur, and apples may have been among the fruits they cultivated. Early cuneiform texts from Mesopotamia also mention strings of dried apples, although textual references to apples are highly unreliable because of the very broad application of the words for fruits. For example, some scholars think that the golden apples of the classical garden of the Hesperides may have been citrus fruits.

Whether the Ur apples came from wild or cultivated trees, it seems likely that apples were cultivated in Mesopotamia by at least the early first millennium BC, as attested by archaeobotanical remains of small preserved apples

Figure 23. Left: two views of an archaeobotanical apple seed. Right: a modern apple seed. I recovered the ancient seed from Tuzusai in the southern Tien Shan Mountains in 2010. Dating from ca. 400 B.C., it is currently the oldest *Malus/ Pyrus* seed ever recovered from this region, the accepted origin of domestication for our modern apple. It morphologically resembles the seeds of the true wild apples that still grow in the region.

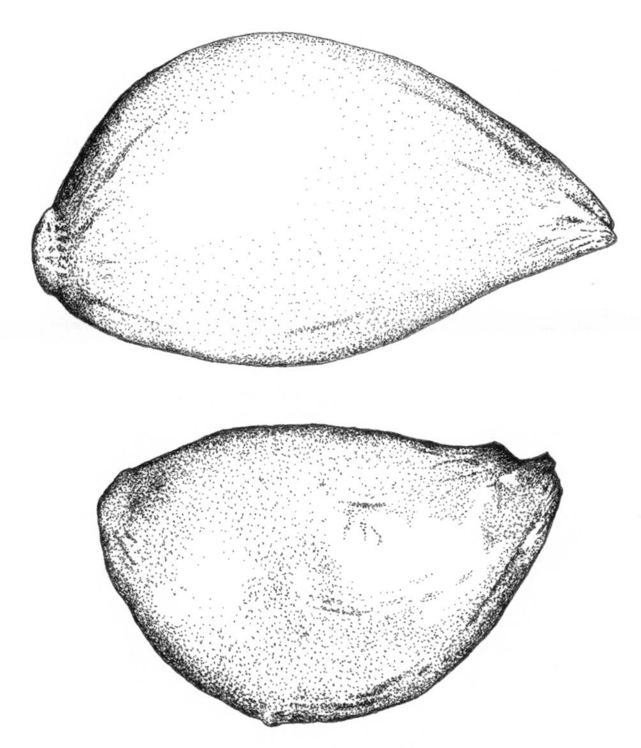

Figure 24. Side views of two apple seeds. Top: a seed from a Cortland apple grown in my backyard. Bottom: a carbonized seed from the ancient site of Tashbulak, Uzbekistan.

from the ancient village of Kadesh Barne'a; however, it is unclear what species they belong to. This site is located on the border between the Sinai Peninsula and the Negev Desert of Israel. A similar cache of desiccated, halved apples at the site dates to the tenth century BC and likely represents a cultivated stock.[69]

By the time of the Roman Empire, the apple was revered from one end of the Silk Road to the other. However, the archaeological evidence of apples in Central Asia before the medieval period consists of one possible apple seed recovered at Gonur Depe in the Murghab Delta of Turkmenistan (2500–1700 BC)[70] and a single apple seed that I recovered from the late-first-millennium BC town of Tuzusai (figure 23).[71] This site is located among wild apple forests, and, despite a lack of evidence, people in the region likely were the first to domesticate true apples. The large seeds found at the Silk Road mining town of Tashbulak, high in the Pamir Mountains of Uzbekistan (figure 24), dating to the late first millennium AD, corroborate the fact that apples were traded along the Silk Road.[72] The settlement sits at an elevation of roughly 2,200 meters, and late spring frosts would have made apple cultivation at the site impractical. However, land at elevations better suited to orchards would have been a few hours' walk away.

CHAPTER TEN

Other Fruits and Nuts

THE *PRUNUS* FRUITS

The genus *Prunus* in the Rosaceae family, which includes plums, peaches, apricots, cherries, and almonds, flourished with the increase of human travel along the routes of the Silk Road. Most of the drupes in this genus, whether wild or cultivated, are sweet and easily dried for storage and transport. The link between cultivated *Prunus* species and the ancient Silk Road goes back well beyond the first millennium BC: dried apricots and cherries have been sold at Central Asian bazaars and carried by caravan traders since prehistoric times. I recovered peach, cherry, and apricot pits at Tashbulak, Uzbekistan (figure 25), and they are all prominent in archaeological reports from medieval urban sites in Central Asia, especially sites excavated during the 1960s and 1970s.[1] Peaches were cultivated across Inner Asia and have been recovered from the Khorezm Oasis in western Central Asia from archaeological layers dating to the fourth or third century BC.[2] As the Silk Road brought distantly related cultivars and wild populations of these fruits into contact, a range of hybrids developed.

Plums

The European plum (*Prunus domestica*) is divided into three clades or subspecies: the European plum (*P. domestica* ssp. *domestica*), the damson plum (*P. domestica* ssp. *insititia*), and *P. domestica* ssp. *italica* (the gages). However, these three species hybridize, and their morphology overlaps considerably.

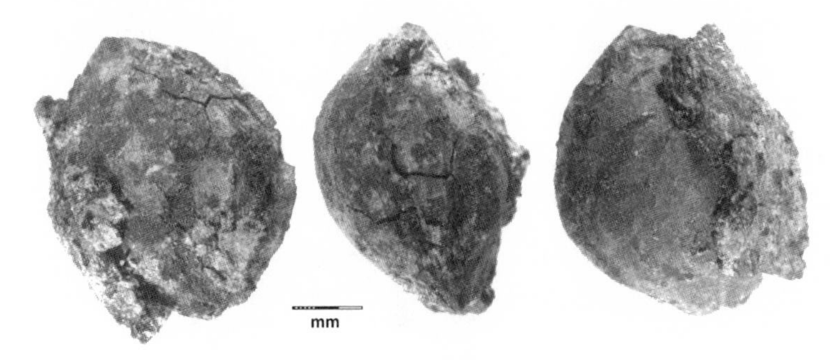

Figure 25. Carbonized apricot pit, with dried, preserved fruit flesh still adhering to the stone, from Tashbulak, Uzbekistan (AD 900–1200).

A fourth plum relative, *P. cerasifera* (sometimes called a cherry plum), is grown in southwest Asia and the eastern Mediterranean today. When these species were first introduced to various parts of Asia is not certain. At least one variety of plum was cultivated in Syria during the early Islamic period.[3] A plum pit recovered from the tenth-century village site of Tell Shheil 1 in Syria still had preserved carbonized fruit material attached to it, illustrating that it was dehydrated before it was carbonized.[4] *Prunus salicina* (Chinese plum) is grown in Syria today and may have been introduced from China; *P. domestica* was likely introduced from the Mediterranean.

The Golden Peaches of Samarkand

The ancient Latin name for the peach is *malum Persicum* or Persian apple, which is not too far from our modern taxonomic name of *Prunus persica*, or Persian plum. This nomenclature attests to the fruit's long history of cultivation in the Persian world. Evidence suggests that the peach originated in China and spread along the Silk Road to Europe via southwest Asia. While Alexander the Great is often credited with introducing the peach to Europe through his Persian conquests, this is likely a fanciful story: he is credited with the introduction of a number of crops to Europe for which he was probably not responsible.[5] However, in the case of the peach, the timing may not be too far off: it was probably brought from southwest Asia to the Mediterranean in the middle of the first millennium BC. Likewise, the earliest trustworthy European reference to the fruit comes from Theophrastus, who places its origin in Persia; he may have seen it while on campaign with Alexander.[6] It appears in Classical myths and art, such as a famous fresco from the ancient city of Herculaneum, depicting peaches and a jar of water, which is thought to have been painted around AD 50 and was entombed by the eruption of Mount Vesuvius in AD 79. The peach is mentioned by Pliny the Elder, who met the same fate.[7] His writings imply that peaches had recently been introduced to Rome. The peach never thrived in India or other parts of South Asia, likely due to the warm climate: peaches need a period of cold dormancy each year to produce a significant crop.

Recently, eight peach pits were recovered from stratigraphic layers eroding out of a hillside near Kunming in Yunnan, China. This find was exceptional because it dates to the Late Pliocene (ca. 2.6 million years ago).[8] These pits are morphologically very similar to some cultivated landraces of peaches that grow in parts of China today. Paleontologists call these fossil peach pits *Prunus kunmingensis*, and they suggest that long before humans began cultivating the peach, its fruits were already large. These fruits, with hard, thick-shelled pits, would have been well adapted to dispersal by large mammals, such as early Asian primates or later Pleistocene megafauna. Early hominids could have played a part in this process of dispersal of the Pliocene fruits.

The *persica* clade across China is highly diverse, especially in the geographic arc from Mongolia to the southern Himalaya. Feral or wild peaches and apricots grow across Central Asia today, from Gansu and Xinjiang to the Fergana valley of Uzbekistan. A close wild relative, *Prunus ferganensis*, grows in the Tien Shan and Pamir Mountains; it is also cultivated in parts of Central Asia.[9] In Linzhi County in eastern Tibet, an area with an especially high number of landrace varieties, there are extant wild trees of *Prunus mira* that are more than one thousand years old.[10]

Most scholars have placed the domestication and original cultivation of our modern peach somewhere in the vast area of northern and western China, likely in the mountainous regions of Tibet and Yunnan.[11] Chinese legend places its origins in the Kunlun Mountains of western China.[12] However, recent archaeobotanical research has shed new light on this question, suggesting that the peach originated along the marshes of the Yangtze River Valley in Zhejiang, China.[13]

Peach pits were recovered from holes dug for storage at the Hemudu site (4900–4600 BC) in the lower Yangtze valley of eastern China, illustrating that people were collecting the fruit from wild trees and keeping the pits in order to consume the seeds within.[14] While the people at Hemudu engaged in early forms of rice cultivation, they relied mainly on hunting and foraging. The archaeobotanical assemblage at the adjacent and contemporaneous site of Tianluoshan indicates that its inhabitants were also low-level rice cultivators and collected wild peaches.[15] Peach pits were also recovered from the slightly older but nearby archaeological site of Kualmqiao (6000–5400 BC), in Zhejiang Province.[16] Other finds of peach pits have been reported at sites of rice-farming peoples in the Longshan (3000–1900 BC) and Yangshao (5000–3000 BC) cultures, all in Henan Province—notably at Yangcun, Shiyangguan, Wuwan, Youfangtou, Xiawu, and Jizhaias as well as Shidao in the Erlitou culture (1900–1500 BC).[17] The Jizhai archaeobotanical assemblage included remains of apricot pits.

Recently a team of Chinese archaeologists compiled a list of twenty-four archaeological sites where peach remains were found, mostly located in the lower Yangtze River Valley and dating to between 6000 and 200 BC. They also noted two finds of peach pits at sites in the Jomon culture of Japan, in the same date range. They explored the morphology of peach pits from

Figure 26. Half a peach pit from Tashbulak, Uzbekistan (AD 900–1200).

Kuahuqiao (6000–5000 BC), Tianluoshan (5000–4500 BC), Maoshan (2900–2600 BC), Bianjiashan (2500–2400 BC), and Qianshanyang (2200–1900 BC), all in Zhejiang Province. By the time of the Liangzhu culture (3300–2300 BC), they note, peach pits were larger and more similar to those of modern landrace varieties. They argue that by 5500 BC, people in the lower Yangtze River region already had a close coevolutionary relationship with peaches.[18]

The peach appears repeatedly in ancient Chinese texts and poetic writings. It was described as early as 1000–500 BC in the *Shijing* (Book of odes).[19] Both peaches and apricots are mentioned in the Donhuang texts from Gansu, an oasis outpost along the Silk Road. Most interesting of all are the finds of stones from both types of fruit in burials from the Sampula cemetery in Xinjiang, 400–100 BC.[20] Peach pits were also recovered from the Silk Road city of Tashbulak (AD 900–1200) in the Pamir Mountains of Uzbekistan, illustrating their importance in Central Asia during the peak of trade through that region (figure 26).[21] Both peaches and apricots had reached the Indus Valley, in northern Pakistan, and Kashmir by the late Harappan period, after 2000 BC.[22]

Like the apple, the peach has been deeply entangled with mythology and legend. In China, especially in the Daoist tradition, it was a symbol of immortality. It was so esteemed in ancient China that the historian Edward Schafer used it in the title of his comprehensive study of exotic imports into the Tang dynastic center, *The Golden Peaches of Samarkand*.[23] In this account he suggests that a golden landrace variety of peaches grown in Central Asia was among the most coveted of the exotic items that moved along the Silk Road during the late first millennium AD. He recounts that Emperor Taizong (r. AD 629–49) was presented with peaches from the people of Kan (Sogdians from Samarkand) that were golden in color and the size of goose eggs. By imperial decree, the pits were planted in the royal gardens. Today there is a wide range of peach landrace varieties in China, ranging in color from tan to yellow to red, and in size from larger than a fist to flat like a donut, such as *P. persica* var. *platycarpa*.

Praised for its sweet fruits, delicate flowers, and beautiful bark and foliage, the peach tree is one of the most frequently depicted plants in East Asian art.[24] Chinese ink-on-rice-paper landscape paintings, a tradition that dates back at least two millennia, often feature a flowering peach tree. Important people in China were often depicted in portraits holding the peach of longevity (*shou t'ao*). The *shou t'ao* was said to have come from a peach tree in the orchard surrounding the jade palace of the Empress of the West, which produced a ripe peach only once every three thousand years. This myth underscores the belief that the peach hailed from the Kunlun Mountains in the west. In a related legend, Sun Wukong, the Monkey King, immortalized in Wu Cheng'en's sixteenth-century *Journey to the West*, was said to have stolen the immortal peach, and Chinese artists often depict a monkey holding a peach.

Among the many examples of peaches in ancient Chinese poetry, Tao Yuanming (AD 365–427) plays on a theme of peaches in a number of poems, most notably "The Peach Blossom Spring," written in AD 421. In another poem, he uses the peach tree as a metaphor for a young bride:

Graceful and young the peach tree stands;
How rich its flowers, all gleaming bright!
This bride to her new home repairs;
Chamber and house she'll order right.

Graceful and young the peach tree stands;
Large crops of fruit it soon will show.
This bride to her new home repairs;
Chamber and house her sway shall know.

Graceful and young the peach tree stands,
Its foliage clustering green and full.
This bride to her new home repairs;
Her household will attest her rule.

(Tao Yuanming, *The Book of Poetry*,
translated by James Legge, 1876)

Peach wood is still carved into figurines and other protective icons in China to repel evil spirits.[25] Intricately carved pits from the peach and Chinese olive (*Canarium album*) were worn as protective amulets. Ornamental peach trees are still commonly planted around China for their flowers rather than their fruit.

Peaches are also a recurring theme in ancient texts from Central Asia and the Persian world. They are mentioned repeatedly throughout the *Memoirs of Babur*.[26] They are also named in the *Ain-i-Akbari*, the accounts of Emperor Akbar written by Abdul Fazal in the sixteenth century, alongside other Central Asian fruits, including apples, pistachios, pomegranates, almonds, quinces, and pears.[27] Their prominence in these texts is supported by archaeobotanical evidence of their importance as orchard fruits. A peach pit was recovered from Diban 5, in northern Syria, from occupation contexts that date between the mid-eighth and the ninth century.[28] A peach pit was also recovered from a twelfth-century hearth at Qaryat Medad, also in Syria.[29] Some of the best-preserved archaeobotanical remains from the Iranian

Plateau were found in the salt mine of Chehrabad, in the Zanjan region of northwestern Iran. Miners working the cave in 1994 came across human remains of Salt Man 1, which were dated to 1,700 years ago. The high salt content in the cave has preserved an astonishing array of archaeological artifacts in wonderful condition, including grains, fruits, and nuts. Among the fruits were peach and apricot pits that date to the Achaemenid period (550–330 BC).[30] They also included remains of Achaemenid watermelon seeds, the earliest clear evidence for this African domesticate in southwest Asia, and remains of *Elaeagnus* pits, fig seeds, grape pips and pedicels, walnut shells, and acorns from the Sasanian period (AD 224–651).[31] This variety of fruits further illustrates the diversity of the orchards and vineyards of southwest Asia from the first millennium BC until shortly before the modern era.

Yellow Apricots

While most of the world's cultivated apricots today are members of the species *P. armeniaca*, a number of other closely related species from Central Eurasia have undoubtedly had a close relationship with humanity for millennia, including *P. brigantina*, *P. mandshurica*, *P. mume*, and *P. sibirica*. The modern domesticated apricot was known to have been grown widely in Armenia, and, as its name suggests, it was long thought to originate from there. Despite this view, the early-twentieth-century botanist Nikolai Ivanovich Vavilov placed its center of domestication in China.[32] Other scholars have proposed different areas of domestication, including India. Although the earliest clear evidence points to an origin alongside peaches in China, all of these scholars may be correct to some degree, as there were likely wild populations of apricot across Asia. The sweet flesh of the apricot would inevitably attract any humans who came across it, and therefore, it might have been brought under cultivation repeatedly in different areas.

Like the peach and many other fruits, the apricot is often said to have been introduced to Macedonia by Alexander, but there is little evidence to lend credence to this myth. Apricots are important in Chinese traditional medicine, and sometimes practitioners of the medical system are referred to in China as "experts of the apricot grove," harking back to an ancient Confucian allegory. A few historians have suggested that the apricot made its way to Iran by the second or first century BC, and to Greece shortly thereafter.[33] The common belief that the tree originated in the Caucasus likely comes from the use of the Latin name *Mailon armeniacon* by Dioscorides and other Classical writers. In Turkey, an old adage implies that nothing could be better than an apricot in Damascus. Pliny the Elder mentions an early-blooming variety of peach that he calls *praecocium*.[34]

Literary references to stonefruit in China may go all the way back to Emperor Yu (2205–2198 BC). Scholars have suggested that oracle bones dating to the Sang Dynasty (ca. 1558–1046 BC) bear the ancient Chinese symbol for the apricot, and there are references to apricot orchards that date to 406–250 BC.[35] Additional literary and historical evidence supports an eastern Chinese origin for the fruit.[36] Archaeobotanical identifications of apricot are hindered by the fact that plums are common in China, especially *Prunus salicina* and *P. mume*, which closely resembles the apricot. At ancient Chinese sites, the apricot is far less common than the peach, but it does appear in the archaeobotanical assemblage from Jizhai in Henan Province.[37] It also appears in burials from the Sampala cemetery in Xinjiang (400–100 BC).[38]

Apricots are a major crop in the Upper Zerafshan region today, and more than thirty local varieties are attested in historical sources.[39] Historical sources also note apricot orchards across the Zerafshan region up to elevations of 2,000 meters. The discovery of seventh- and eighth-century apricot pits, along with a variety of other fruit remains, in the fortified site of Mugh in northern Tajikistan suggests that they were a prominent part of the landscape in the past there as well.[40] Other contemporaneous finds of apricot pits in the same region date to roughly the same period.[41]

Today, most of our cultivated apricot trees are grafted onto hardier peach rootstocks, and self-pollinating varieties have been developed. However, in antiquity, the tree was far more particular about its preferred soil conditions, nutrients, rainfall, and proximity to another tree for pollination. While these factors may have slowed its spread, the historical and archaeological evidence shows that it was well known across the ancient world by the first millennium BC and likely much earlier.

Most landrace varieties of apricot tend to be more cold tolerant than peaches. However, because they bloom early in the spring, crops are often threatened by late freezes and night frosts. Feral apricots grow across Central Asia today, but it is unclear whether they are a legacy of the Silk Road or of Soviet agricultural campaigns.

Cherries

There are dozens of species of wild cherries across Eurasia, several of which are native and would have provided easily accessed snacks along the Silk Road. However, most modern cherry cultivars are descended from *Prunus avium*, the sweet cherry or bird cherry (a name also applied to *P. padus*), or, to a lesser extent, from *Prunus cerasus*, the sour cherry. *P. avium* originates in West Asia and is common in the wild from southwest Asia to the Himalaya. Common wild cherry species in Central and Western Asia include the sweet cherry, the dwarf or ground cherry (*P. fruticosa*), and the sour cherry.[42] There are several other economically significant species in East Asia, including the Japanese apricots (*P. mume*), the Chinese cherry (*P. pseudocerasus*), and the bush cherry (*P. tomentosa*).

In his survey of the wide range of landrace varieties of cherry that were cultivated in ancient Rome, which he refers to as *cerasus*, Pliny compares their sweetness. He claims that Lucius Licinius Lucullus brought the sweet cherry to Rome from the Pontic region after the conquest of Mithridates VI in 74 BC. In Pliny's account, Lucullus became acquainted with the fruit either after the Battle of Tigranocerta in modern-day Armenia in 69 BC or during the Third Mithridatic War (73–63 BC).[43] Alphones de Candolle and subsequently many others have pointed out, however, that according to the archaeobotanical evidence, many forms of cherries had been consumed in Italy and across Europe for millennia. While it is possible that Lucullus brought back a specific landrace variety from the Caucasus, he did not introduce cherries to Italy.

Preserved cherry pits have been recovered from archaeological sites across Central Asia; however, given the abundance of wild species in the region, there is no particular reason to believe that the recovered pits are from cultivated trees. In fact, some of the pits, such as those from Tashbulak in Uzbekistan (AD 900–1200), seem to be from a small local variety.[44] Cherry pits were also recovered at the site of Sarazm (3500–2000 BC), not far from Tashbulak. In addition, cherry pits have been recovered from sites in southern Central Asia, such as the village of Adji Kui in the Murghab region of Turkmenistan, dating to the second millennium BC.[45]

MELONS

Studying the cultivated cucurbits—including melons, cucumbers, and squash—in Asia is almost as hard as studying the cultivated legumes. Several species in the Cucurbitaceae family have been cultivated in China for millennia. Most of them, however, never spread along the Silk Road to Europe and were overlooked when colonial plant hunters were tramping across the globe in search of interesting species to introduce to Europe. For example, the bitter melon or bitter cucumber (*Momordica charantia*), the wax gourd or winter melon (*Benincasa hispida*), and the snake gourd (*Trichosanthes cucumerina*) are uncommon in Europe and the Americas. Furthermore, most Euro-Americans know the luffa gourd (*Luffa acutangula*) only as the funny sponge that hangs in their shower stalls, and the bottle gourd (*Laginaria siceraria*) as a whimsical gourd for making ornamental birdhouses. While all of these species have long histories in East Asia, the only cucurbit that seems to have played a significant role on the Silk Road is the common melon or honeydew (*Cucumis melo*).

Despite the large number of domesticated species and frequent historical references to melons and gourds, there are essentially no archaeobotanical data for early cucurbit cultivation in East Asia, partly because very few seeds have been preserved. There is an early published report of preserved seeds at a fourth- and third-millennium BC site at the mouth of the Yangtze River, but at least one scholar has noted that the site is disturbed and unreliable.[46] The common melon did not appear in China until the fifth or sixth century AD.[47] On the basis of early Chinese literary sources, some historians claim that it did not make its way to China until the eighth century AD.[48]

While scholars in the past have argued for an African origin for the common melon,[49] more recent genetic studies have suggested an origin somewhere on the continent of Asia (or, oddly, Australia). Recent genetic work has supported the generally accepted view that the cucumber (*Cucumis sativus*), a close relative of the melon, was domesticated in India.[50] Furthermore, geneticists seem to have identified the wild progenitor as *C. sativus* var. *hardwickii*.[51] The debate over the exact origin of the melon continues; while some scholars still hold to the idea that it came from Africa, the most likely center of origin is southwest Asia. Like so many of today's familiar crops, it probably originated in the Fertile Crescent, but much later than the founder crops.

There is a remarkable diversity of cultivated melons. The monophyletic clade of what are commonly known as muskmelons is divided into three main subclades on the basis of skin texture. The subspecies *C. melo* ssp. *reticulatus*, whose members have reticulated structuring on the skin, includes the familiar American cantaloupe. This subspecies is not very popular in Central Asia today and probably was not present on the Silk Road. A lesser-known subspecies, *C. melo* ssp. *cantalupensis*, with rough or warty skin, includes the Persian melon as well as the smooth-skinned European cantaloupe and has a long history in southwest Asia. The third subspecies is the most diverse and has long been known along the routes of the Silk Road: the Asian melon or honeydew, *C. melo* ssp. *inodorus*. These melons have been cultivated across Central Asia and China for millennia, and a bewildering number of landrace varieties have developed, including the green-and-yellow-striped Hami melon from Xinjiang, an ancient variety that has received a lot of attention from breeders.

Farmers across all of Central Asia grow melons alongside other crops. This practice has led to the development of countless local varieties. An autumn traveler entering any village from Kazakhstan to Turkmenistan would have been entreated by produce vendors to try the local melons. Growers take great pride in their local varieties, a legacy of millennia of cultivation and the spread of their seeds along the great Silk Road.

I recovered a few preserved carbonized melon seeds from Tashbulak during excavations in 2015. The seeds, dating to between AD 800 and 1100, were found in a midden in the central square of the village.[52] It is likely that the melons were brought up to the village from nearby valleys. Melon seeds dating to the fourth or fifth centuries AD were also recovered from the Khorezm Oasis in western Central Asia, at the site of Kara Tepe.[53]

OTHER SILK ROAD FRUITS

A number of other plant foods were transported and cultivated along the routes of the Silk Road. These include Russian olives (*Elaeagnus* spp.), olives (*Olea europaea*), figs (*Ficus* spp.), pomegranates, dates (*Phoenix dactylifera*), hawthorn berries (*Crataegus* spp.), jujubes, persimmons (*Diospyros* sp.), sea buckthorn berries, hackberries, and mountain ash berries (*Sorbus* spp.).

Some of these fruits were important in Central Asia in the past but never spread much farther west. For example, Russian olives are cultivated in the

Figure 27. Melon vendor in Samarkand, 1911. Photo by Sergei Mikhailovich Prokudin-Gorskii, using an early color-plate method. Library of Congress Prints and Photographs Division, Washington, DC.

foothills of Uzbekistan today and can often be found in markets there; however, the tree appears in America only as an invasive ornamental. The fruit has a mild flavor. Like its namesake, the olive (to which it is not related), it has a large pit. It was cultivated as far west as the Khorezm Oasis along the southern end of the Aral Sea as early as the fourth or third century BC.[54] Pits of Russian olives dating to the Sasanid period, in the middle of the first millennium BC, have been recovered from the Chehrabad salt mines of north-

Figure 28. Melon vendor's stand at the fruit market in Bukhara, 2017. Photo by the author.

western Iran.[55] Wild shrubs of this species grow across the Iranian Plateau and adjacent regions today. Likewise, sea buckthorn, although it has gained some popularity as a nutraceutical in America, is harvested primarily in Russia and Central Asia, where it is often fermented into alcohol. Many of these fruits have left only the faintest of historical and archaeobotanical trails, and it is not clear how prominent they were or how far back in time their story on the trade routes begins.

Pomegranates

The pomegranate has been prominent for millennia in culinary and mythological traditions across southwest Asia, the Mediterranean, and South Asia, including India. It was traveling along the Silk Road (or the Spice Routes) to East Asia by the early first millennium AD. The pomegranate was spread far and wide by both Turkic and Persian travelers: the word for pomegranate in Armenian, Bulgarian, Maldivian, Punjabi, Hindi, Tajik, Uzbek, and Kazakh is related to its Farsi name, *anar*, and the Turkic name, *nar*.[56] Its ancient Roman name, *Malum punicum*, or Punic apple, implies that it was brought to Rome from Carthage. The fruit is anatomically quite different from the other fruits discussed in this chapter, with individual seeds surrounded by sweet red arils inside the fruits. Because the shrubby plant can survive in rather arid locations, it is widely grown across southwest Asia today.

The pomegranate has long been featured in art and literature. It appears repeatedly in the Old Testament and in Greek, Roman, and Persian art. It has held religious significance across the ancient Mediterranean as well as in Asia. It also spread east: the ancient Wei Dynasty book *Qimin yaoshu*, likely completed around AD 544, suggests that pomegranates had recently been introduced from Central Asia.[57] On reaching China, somewhere between the late third and the fifth century AD, the fruit acquired a symbolic meaning in Buddhism.[58]

In Greek myth, the six pomegranate seeds eaten by Persephone, which condemn her to spend half of every year in the underworld before emerging in the spring, associate the fruit with annual cycles of growth and fertility. The juicy red fruit represents everything untamed in nature, from lust and carnal desire to death and rebirth. Zoroastrian tradition places the fruit on the Nowruz festival table as a symbol of longevity; Isfandyar, the Persian warrior, drank the juices in preparation for battle. The latter practice spread across the Silk Road in the thirteenth century and was adopted by the Turco-Mongol conqueror Timur. The fruit is still laid at his tomb in Samarkand today.

Dates and Figs

The date or phoenix palm is another southwest Asian fruit that may have been brought into China by Iranian merchants during the first millennium AD. However, wild palms are native to Yunnan and other areas of South Asia. Since the tree is not well adapted to many areas of China and clearly could not have survived the cold winters in northern latitudes or the high mountains of the Silk Road, it must have made its way into China along southerly routes. However, because the fruits are high in sugar and easily dried for long-distance transport, they merit mention here. The only archaeobotanical reports of date pits along the Silk Road come from one Soviet-era find at the Bazar Dara mining town, dating to about a millennium ago in Uzbekistan. Similarly, figs (*Ficus* spp.) cannot grow at the high altitudes of Central Asia, but today they are cultivated at lower elevations as far north as Uzbekistan. Of the several southwest Asian fig species found in subtropical China, the most prevalent today is *F. carica*, the same species that is cultivated across the Mediterranean.[59] The fig prefers warm, dry climates and clearly spread to Pakistan and northern India before making its way to southern China. Historians have suggested that the tree did not reach China before the eighth century AD.[60]

Capers

Wild *Capparis spinosa* grow across a large stretch of Central Eurasia, in arid regions and often on disturbed soils. The short, spiny plant, with its gorgeous flowers, is often found along the edges of abandoned human settlements. While many Euro-Americans are unfamiliar with the fruits, most are familiar with the sharp taste of the pickled caper buds, which are an important part of Arabic, Greek, Italian, Moroccan, Spanish, and Turkic cuisine. The flower buds are harvested before they open and pickled in brine to cut their natural bitterness. When the flowers are allowed to bloom, the fruit is often pickled in a similar fashion.

Understanding the origins and cultivation of this plant is difficult because it grows wild from the Arabian Peninsula to Russia, often in patches dense enough to make cultivation unnecessary. On several different trips to Uzbekistan, I have seen women collecting the fruits from arid steppe regions, and they are also collected by Uighur women in Xinjiang. The seeds, which are found in archaeological sites from an equally broad region, likely represent collections of the fruits from the wild.

Persimmons

The most important commercial species of persimmon today is *Diospyros kaki*, which originated in Japan, eastern China, and Korea. There is no evidence that the persimmon traveled into Central Asia in antiquity, likely because of its highly perishable nature. However, the persimmons of China display amazing genetic diversity, and various species in the *Diospyros* genus were consumed across the Northern Hemisphere in antiquity. In North America, the species *D. virginiana* is still collected from wild stands in the southeast and central states. It is also found in archaeobotanical assemblages predating European contact. In addition, the date plum (*D. lotus*) has been cultivated or collected from the wild across much of southern Eurasia, from China and India to the Mediterranean. A few persimmon remains have been found at Han archaeological sites in China, and there is a long tradition of cultivating *D. kaki* from Yunnan to Beijing.[61]

Jujubes

Another fruit that enjoyed some prominence on the early routes of the Silk Road is the jujube. Again, however, little archaeobotanical evidence of its presence in Central Asia has been recovered. These trees were grown over a large area of the ancient world. They have fruits very similar to Russian olives, with a mild flavor and a large pit inside. The fruits are often dried for transport and long-term storage; they are sometimes rehydrated by boiling in tea. The most widely cultivated species is *Ziziphus jujuba*. According to some sources it was originally domesticated somewhere in South Asia, probably India, very early in prehistory.[62] From there it likely spread along more southerly routes to the Mediterranean. This crop is not as common in European markets as it once was, but it is still cultivated in small enclaves in the Mediterranean region and is very popular in South Asia.

Although Vavilov initially entertained the possibility that the fruit was first domesticated in Central Asia, it probably originated farther south. It is mentioned in early written sources from India and China and it is quite common in Classical sources. Theophrastus, Dioscorides, and Pliny were probably acquainted with *Z. jujuba* or *Z. lotus*. Pliny claims that the common jujube was introduced to Rome from Syria by the consul Sextus Papinius toward the end of the Augustan period (27 BC–AD 14); but, like most of Pliny's claims of this sort, this was probably just rumor. Most likely, both *Z. jujuba* and *Z. lotus* were well known across the Mediterranean in the classical period.

The latter species, commonly grown across the arid regions of southwest Asia, was probably the lotus tree mentioned by Herodotus and Theophrastus.[63] It is also possible that the intoxicating lotus fruit mentioned in the *Odyssey* and by Herodotus was a magical form of *Z. lotus*. The sacred lotus tree next to the Temple of Vulcan in Rome, described by Pliny, was likely a *Z. lotus*. This species and another southwest Asian species, *Z. spina-christi*, are mentioned in many Arabic and Islamic sources, including the Quran, as lote-trees.

Ancient people across the Old World likened the jujube fruit to the date when dried. Ancient literary sources from southern Europe refer to the dried fruit as the Chinese date, and ancient Chinese sources call the dried date-palm fruits Persian jujube.[64]

When ripe, the jujube fruit is plump, brown, and sweet, but it is usually sold dried, with a reddish, wrinkly skin. In this form it was readily transported along the Silk Road. Desiccated jujube fruits were found along the routes of the Silk Road in burials in Turfan at the Astana site dating to the Tang Dynasty.[65]

Because the shrubby jujube tree does not require a lot of water and can withstand cold, dry winters, it was well suited to many parts of southwest and Central Asia. Another species in the genus, cultivated across India and South Asia, is less tolerant of frost. Wild jujube trees still grow across northern China, and hundreds of landrace varieties can be found. By the first century AD in China, scribes were praising the jujube orchards in Shansi Province, and Tang Dynasty accounts even mention a jujube wine.[66]

Hawthorn Apples

There are several species of *Crataegus* shrubs or trees in China that are cultivated or gathered from the wild, and several more in Central Asia. They are common shrubs or low-growing trees in the shrubby forests that likely covered much of the Central Asian foothills before the second millennium BC. Today they exist in small remnant patches in the foothill ecotone. Hawthorn stands, despite having long thorns and being particularly resistant to herd-animal defoliation, are restricted to rich valleys and areas less impacted by browsing. They are some of the only woody species able to withstand heavy pastoral grazing in Central Asia. They produce sour-sweet fruits (small pumes) that are high in vitamin C.

A few hawthorn species have been domesticated, and two are still cultivated today: *C. pinnatifida* and *C. hupehensis*.[67] *C. pinnatifida* is often skewered and glazed with sugar, like a candied apple at an American carnival. These candied haw apples can be found at many night markets across China, notably in Beijing. *Crataegus* (and *Rosa* spp.) seeds have been found at several archaeological sites in Central Asia, including Tuzusai (410–150 BC) in Kazakhstan,[68] Tashbulak (AD 900–1200) in Uzbekistan,[69] and Adji Kui (ca. 1900 BC) in Turkmenistan.[70] Kazakh herders traditionally collected *Crataegus* hips in the autumn; these shrubs also grow on the steppe proper and provide a nutrient-rich food available seasonally across Inner Asia.[71]

Other Fruits

The ecology of the foothill zone of Central Asia has changed dramatically since the introduction of farming and pastoralism. At one time, much of it was covered in short-growth forests of shrubby trees. Historically, these forests, rich in wild fruits, were important in human economies from the Altai to the Pamir. They were dominated by *Viburnum opulus* (common viburnum) in the Dzungar Mountains and *Hippophae rhamnoides* (sea buckthorn) along alluvial deposits and riverbanks in the Tien Shan. Several wild rose (*Rosa* spp.) species grow across the Semirech'ye region and in most of the ecological zones. Two species of *Elaeagnus*, or Russian olives, grow throughout Semirech'ye and range south into the Pamir, and, like species of *Rosa*, are present from the mountain forest regions down to the edge of the arid steppe. *E. angustifolia* is more common in the mountain forests, while *E. oxycarpa* is abundant across the foothills. Several species of raspberry (*Rubus*) share the same environmental range; four were collected by historical populations of traditional peoples living in the mountains of Kazakhstan.[72]

These forests have, or had, several species of wild cherry (*Prunus* and *Cerasus* clades), which were collected in Kazakhstan and farther north as well.[73] At least seven species of *Crataegus* grow in the foothills of the eastern Kazakhstan mountain ranges; several are known to have been collected from the wild in the Almaty area.[74] The best known of these shrubby trees are wild apples, represented by two species in the mountains, *Malus sieversii* and *M. niedzwetzkyana*.

These wild fruits and nuts represented important food resources. Early Russian and Western European explorers mention gathering cranberries (possibly *Vaccinium opulus*) in the northerly forests by local people.[75] Some accounts claim that *V. microcarpus* and *V. palustris* were harvested in Kazakhstan.[76] Crowberries (lingonberries or fox berries, *V. vitis-ideae*) were collected in the Altai Mountains.[77] Bilberries (*V. myrtillus*) were collected by mobile pastoralists farther north and eaten raw, boiled, or mixed with cream, or milk (fresh or fermented). Some of these explorers described people collecting cloudberries (*Rubus chamaemorus*) in the Altai Mountains.[78] Eleven species of *Ribes* grow in Kazakhstan,[79] and ethnographic observations from the nineteenth century note the collecting of redcurrants and blackcurrants (*R. vulgare* and *R. nigrum*).[80]

These mixed shrubby forests likely covered much of the Kopet Dag Mountains as well as the foothills of more distant mountains, such as the Iranian Plateau and the Pamir Mountains. Here they included wild pistachio, wild almond, *Elaeagnus*, *Crataegus*, and cherry trees. At Sarazm in Tajikistan, archaeobotanical remains of foraged nuts and fruits include capers, hackberries, wild pistachio, Russian olive, cherry, rose hips, sea buckthorn, and wild almonds.[81] Capers were also reported in the macrobotanical assemblage at Jeitun.[82] Fragments of wild hackberry (*Celtis* sp.) pit were found in Chalcolithic layers at Anau[83] as well as Tashbulak (900–1200). *Prunus* sp. and hawthorn stones found at Adji Kui may have come from tree stands that were protected and possibly maintained nearby.

NUTS

The wild forests of the Central Asian foothills were populated by numerous nut trees. Most of these forests have been lost, and today the foothill zone is composed primarily of dry grass fields, used for herd-animal grazing. But the legacy of these forests lives on today around the world.

While almonds (as well as peach and apricot pits), pistachios, and walnuts were likely the main nuts on the Silk Road, it is possible that other nuts were collected and cultivated, such as pine nuts and chestnuts. Pine nuts (*Pinus* spp.), collected from the wild, are still an important food source for people living at high altitudes in Central Eurasia. Several species are sold in markets in western China; two (*P. armundii* and *P. yunnanensis*) are of particular prominence in markets in northern Yunnan. Shell fragments of pine nuts have been recovered from fifth- or fourth-millennium BC sites in the Yangshao culture along the Yellow River.[84] One species, *P. bungeana*, was cultivated around temples and is famous for being grown in Beijing at the Pine of Seven Dragons temple.[85] Shells of *P. yunnanensis/tabulariformis* were recovered from Kyung-lung

Mesa (AD 220–880) in eastern Tibet, a project I was involved with.[86] *P. armundii* nuts are commonly collected today and can be purchased in the markets.

Nuts from *P. pinea* are commonly collected in the eastern and central Mediterranean today. In addition, the seeds of the chilgoza pine (*P. gerardiana*) are collected in Afghanistan and Pakistan, and other species are collected in Korea (*P. koraiensis*) and Mongolia (*P. sibirica*).

There is no evidence that the chestnut (*Castanea* spp.) was cultivated along the Silk Road; however, there are roughly a dozen species in the genus, which is native to North Africa, North America, Europe, and Asia, and they tend to draw the attention of humans wherever they grow. Most Americans and Europeans today are familiar with the European chestnut (*C. sativa*), and the now-endangered American species is featured in Christmas songs. The Chinese chestnut (*C. mollissima*) is commonly cultivated across China. While there is no reason to assume that either the European or the Asian lines ever crossed the mountains of Central Asia, there is a long history of human interaction with this genus on both ends of the great Silk Road.

Pistachios

The pistachio, a member of the Anacardiaceae family that also includes the cashew, has six species native to West Asia, three of which have had a particularly close relationship with humans: *Pistacia vera*, *P. terebinthus*, and *P. acuminate*. Another species, *P. lentiscus*, is used as a source of a resin called mastic. Pistachios are widely grown across Western Asia and the eastern Mediterranean, notably in Iran, Afghanistan, and Uzbekistan to the north. Historical sources from China imply that the pistachio was introduced to China during the Tang period.[87] Strabo claims that when Alexander first entered Bactria (southern Central Asia), the only tree he saw was the shrubby terebinthus, which is also discussed by Theophrastus.[88] Many other classical writers mention the terebinthus tree growing in Syria or parts of southwest Asia, where the nuts were collected from the wild.[89]

The commercial variety known worldwide today is *P. vera*, a species that originated in Central Asia. The earliest finds of this species in southwest Asia date to Hellenic times, with the exception of two shell fragments from Tepe Yahya in Iran.[90] However, Bronze Age remains have been identified at Djarkutan and Gonur Depe[91] and from Chalcolithic layers at Sarazm in Tajikistan.[92] I have also identified shell fragments at Tashbulak.[93]

Walnuts: The King Nut

A recent genetic study of the Persian or English walnut (*Juglans regia*) suggests that its history may be as deeply intertwined with the Silk Road as is that of the apple. Based on pollen studies and modern and historical distribution records, it seems clear that isolated populations of *Juglans* were dispersed across the Hindu Kush and Pamir Mountains, becoming increasingly isolated with the desertification of regions such as the Kopet Dag.[94] Some of the earliest well-attested remains of walnut shells were recovered from Kanispur (3100 BC), in the Kashmir Valley of Pakistan, illustrating their history in the foothills of eastern Central Asia.[95] A recent study suggests that the current dispersal of walnut genetics across Asia is largely a product of human exchange along the Eurasian trade routes. The researchers, who attempted to correlate linguistic data with population genetics, suggest that linguistic barriers mark walnut genetic barriers; conversely, common human cultural areas facilitated intermixing of walnut genetics.[96]

While there are a number of wild species of walnut (*Juglans*) in China, *J. regia* ranges from the foothills of western China all the way to the Caucasus (and the Carpathian Mountains, as it is sometimes called the Carpathian walnut). Walnuts in China and Central Asia display great genetic diversity because they are often grown from seed rather than through cloning, as in Europe. Asian walnuts tend to be much smaller and rounder than European specimens, though some varieties are large and oval, and their shells may be either thin or hard, smooth or rough. Because the *Juglans regia* is genetically compatible with certain wild relatives in China, such as the heart nut (*Juglans ailantifolia*), it is plausible that some of this diversity is due to gene flow between species.

Medieval Arab geographers noted that the walnut was grown only in the more temperate northern regions, notably in the mountainous zones.[97] While exploring Russian Central Asia in the 1800s, Henry Landsell noted the trees of the Zerafshan valley, specifically wild hackberry and walnut groves in the mountain foothills.[98] Similarly, James Fraser, in the early nineteenth century, noted walnut forests as well as pistachio and almond forests in the Central Asian foothills, from northern Iran up to the Zerafshan region.[99] In the early twentieth century, Aurel Stein observed walnut forests in the eastern stretches of the Central Asian mountains, and he discusses the planting of walnut trees at the oases of northwestern China in volume 1 of *Ancient Khotan*.[100] Early floras of Central Asia note that walnut forests were prevalent at intermediate altitudes (1500–2800 meters) and that they grew along the hills around the Fergana Valley. Grave offerings of walnuts were recovered from the Tang Dynasty tombs at the Astana cemetery in the Turfan region of Xinjiang, along with dried jujubes, grapes, and pears.[101] Shell fragments from walnuts dating to about a thousand years ago were also recovered at Tashbulak. Walnut shell fragments and at least one whole walnut were also recovered from the salt mine of Chehrabad in Iran, dating to the early first millennium AD.[102]

One example of the admixing of walnut genetics along cultural lines is represented in nuts from eastern Uzbekistan and the Fergana Valley, which, as the geneticists note, "indicate the exchange of *J. regia* among Turkic communities that lived between Tashkent and Samarkand where the northern and central routes of the Northern Silk Road converged."[103] The researchers argue that the trade routes across Asia were directly responsible for walnut dispersal and genetic mixing, starting in the Achaemenid period. If these conclusions are correct, the long-term human manipulation of the foothill forests of Central Asia may have played a direct role in the domestication of walnuts, which were bred and dispersed across a broad swath of Asia in antiquity.

Literary and linguistic evidence for the origin of walnuts points to southwest Asia. According to linguists and historians, the word used by Pliny and Dioscorides to refer to the walnut translates to "Persian nut" or "nut of the king of Persia"; in addition, one old Chinese name for the nut has Sanskrit roots.[104] The historians also note that the earliest recorded Chinese name for the walnut translates as "peach of the Hu"—Hu being a Sogdian family name referring to peoples of Central Asian and Iranian origins.

Ancient Chinese literary texts suggest that the walnut was introduced to China from either Tibet or southern Central Asia during the first millennium AD, making its way along the routes of the Silk Road from the mountain spine of Eurasia. A text from the Jin Dynasty (AD 265–420), the *Book of Jin*, notes that eighty-four walnut trees lined the imperial park.[105] If we accept that these trees were the common walnut and not a native East Asian species, then it is clear that they were known at least as a novelty in the center of the empire. However, over the following millennium and a half, walnuts became a common crop across much of China, and a bewildering range of landrace forms have sprung up.

Almonds

While most Euro-Americans today eat only the flesh of peaches and apricots, in East and Central Asia people also eat the seed inside the pit, which resembles the fruits' close relative, the almond. Many Americans think of these pits as poisonous. Indeed, apricot pits contain prussic acid or hydrogen cyanide, and consuming a large quantity, especially of the bitter or more toxic varieties, can be fatal. Peach stones found in ancient pits at the Hemudu site in eastern China suggest that people were collecting the stones to process for eating at least six millennia ago, although it is not clear how they were processed.[106] Finds of nuts from *Quercus*, *Lithocarpus*, and *Cyclobalanopsis* in pits from Hemudu and other closely related sites suggest that the pits may have been used to leach out toxins from both acorns and peach pits. Other methods of detoxifying *Prunus* seeds in Asia include fermenting, grinding, and boiling.[107] Some landrace varieties of apricots have been bred more for their seeds than their fruit; these slightly bitter seeds are popular in markets in Central Asia.

Modern almonds have lower toxin levels than the seeds of other *Prunus* species. Wild almonds grow along the foothills or in short-growth forests from Central Asia to the Caucasus and from southwest Asia to northern Central Asia. They were likely an important wild food for people before farming and spread to the Mediterranean very early. There are two primary cultivated forms, bitter and sweet, which differ in their levels of the glucoside amygdalin, a precursor to benzaldehyde and cyanide. Millennia of selection for the sweet form and for larger seeds led to domestication. They may have been known in China by the ninth century AD: several scholars reference the accounts of an Arab merchant who purportedly traveled to China and noted in his logs that almond trees grew in the Tang center of Chang'an.[108] However, these early accounts are not reliable sources of information on almond consumption, because it is easy to mistake shelled apricot pits for almonds. Almonds were also cultivated in northern Pakistan and Kashmir after 2000 BC, along with fruits like peaches, apricots, grapes, and hackberries.[109]

CHAPTER ELEVEN

Leafy Vegetables, Roots, and Stems

LEAFY VEGETABLE CROPS

To most people, "leafy greens" means cabbage, kale, or lettuce. During the time of the ancient Silk Road, however, a wider range of leafy vegetables was available, many of which have been forgotten. Hundreds of wild and domesticated relatives of lettuce (*Lactuca* spp.) and cabbage or broccoli (*Brassica* spp.) were grown across the Old World.

In Asia, the range of diversity within the cabbage clade is even wider. The species *B. oleracea* alone includes kohlrabi, cauliflower, savoy cabbage, European cabbage, broccoli, brussels sprouts, and kale. Broccoli is a variety that has been bred for its large inflorescences over hundreds of human generations; brussels sprouts are a variety of the same plant that has been selected for its leaf buds. When the ancestor of European broccoli spread eastward to China, a different suite of selective pressures yielded *kai-lan*, also known as Chinese broccoli, which has large, flat leaves and a small floral structure. Another species, *B. juncea* (mustard cabbage), is often grown in China but also originated farther west.

The enormous morphological variation within this clade has left many taxonomists stumped over how to classify the multitude of East Asian species, which have an equally impressive and bewildering diversity of common names.[1] Little is known of their origins, and because of the perishable nature of leafy crops, preserved examples are rarely found at archaeobotanical sites. There is no clear historical evidence that *B. oleracea* made its way to China in early antiquity, but various forms were known throughout the Islamic world. Identifying a center of domestication for *B. oleracea* is one of the last great mysteries in plant domestication, because its many forms have developed from various evolutionary offshoots over the past few millennia in regions from southwest Asia to northern Europe. However, recent genetic work is elucidating this narrative.

Genetic studies of the Brassiceae tribe have shown that it diversified roughly 28 to 16 million years ago, and subsequently many species underwent multiple hybridization, which led to the whole genome of the plant becoming duplicated (the same rapid process of domestication described for the polyploid wheats in chapter 7).[2] There are six major species of domesticated *Brassica*, three diploid and three tetraploid. The most distantly related of the diploid cruciferous crops, black mustard (*B. nigra*), was domesticated in North Africa. The other two diploid species, the cabbages and broccolis (*B. oleracea*) and turnips (*B. rapa*), were domesticated in southwest Asia. *B. oleracea* spread mainly westward from the Fertile Crescent, resulting in the abundance of crops mentioned above. *B. rapa* spread both eastward, resulting in bok choy and Chinese cabbage, and westward, producing turnips.

Subsequent hybridizations between the diploid species resulted in three genetically isolated species of domesticated tetraploid crops. *B. juncea* (Indian mustard or mustard cabbage) is likely a cross between *B. nigra* and *B. rapa*, which also originated in southwest Asia.[3] *B. napus* (which includes rapeseed, canola, and rutabagas) appears to be a cross between *B. rapa* and *B. oleracea* that emerged about 7,500 years ago as the result of more chromosome doubling. It spread into China, where rapeseed is widely cultivated today, and into Europe.[4] Finally, *B. carinata* (Ethiopian mustard) is restricted to Ethiopia and Kenya.[5]

In China, Asian leafy *Brassica* species, predominantly *B. rapa* (including bok choy, Chinese kale, and celery cabbage), became popular only in the Tang Dynasty. For millennia before that, another leafy vegetable prevailed in various forms: the leafy green mallow, *Malva verticillata*. This plant was once cultivated across all of Asia and parts of Europe and North Africa.[6] Although largely forgotten, it is still cultivated on a limited scale, especially on the Sichuan Plains in central China, where it is often grown as a low-investment crop along field edges.[7]

Mallow is one of the world's few domesticated biennial or perennial crops. A perennial habit means that the plant needs little tending beyond an occasional weeding. It does not need to be resown after harvesting; seeds are not saved, and it serves additional purposes, such as erosion control at the edges of rice paddies and perennial field demarcations. In addition to being highly nutritious, mallow is powerfully mucilaginous when cooked, like its close relative, okra (*Abelmoschus esculentus*). People in the Sichuan Plains make a slimy soup from it, which, like all Sichuan cuisine, tastes amazing. In 2010, in a small rural town south of Chengdu, I asked for a bowl of the soup at a restaurant. The waitress took me to the field behind the restaurant to pick some. The mallow was essentially allowed to grow wild along the field edge; it was heavily trampled, and little apparent care was given to growing it.

Because leafy crops are rarely preserved in archaeological sites, and their seeds were often not saved for replanting, paleoethnobotanists know very little about their roles in cuisines of the past. Carbonized *Malva* seeds have been recovered from several sites in Central Asia and are abundant at sites like Begash and Tasbas in eastern Kazakhstan[8] and Tashbulak in Uzbekistan.[9] However, interpreting their presence in the archaeobotanical assemblage is complicated. Most of the wild seeds from small herbaceous plants found at these sites were contained in herd-animal dung, which was burned as fuel, carbonizing the seeds.[10] Two species of mallow, *M. neglecta* and *M. sylvestris*, both grow in abundance in the well-watered valleys of Central Asia and were undoubtedly consumed by herd animals. Botanical observations near Tashbulak in 2015 show that in heavily grazed high-elevation meadows, wild *Malva* plants are some of the most persistent herbs and rapidly colonize abandoned herd-animal pens. As a result, we cannot determine whether wild *Malva* seeds found at prehistoric sites across Central Asia are from human or animal food. However, we can start to reconstruct the role of this crop through historic texts and ethnohistoric sources.

According to Egyptologists, a wild form of *M. parviflora* may have been eaten in Egypt as far back as the Late Paleolithic. This early wild food may have been the progenitor for the cultivated *M. parviflora* that is grown in Egypt today.[11] Ancient Chinese, Turkic, and Classical sources describe the use of mallow greens and suggest that they played an important culinary role before the arrival of cabbages. Remnant populations of secluded farmers still grow the crop in parts of East and South Asia as well as in northern Africa and parts of southwest Asia, suggesting that it may have been cultivated more widely in the past.

Ethnohistoric accounts from the first century AD attest to both cultivated and wild varieties of *M. sylvestris* being eaten from Egypt to Rome and throughout Asia. It appears in a number of Classical texts, though we can assume that it had declined in prominence by the Classical period. Dioscorides discusses cultivated *M. sylvestris* in book 2 and wild *Malva* in book 3.[12] Pliny the Elder uses the name *malva* over two dozen times in the *Naturalis Historia*.[13] It is also mentioned in the *Apicius* cookbook as a garden vegetable. *M. parviflora* is still cultivated as a potherb in parts of southeast Asia and North Africa, and it is sold in markets in Egypt.[14]

Some historians claim that *M. sylvestris* was once among the most important vegetable crops in China.[15] Mallow (presumably *M. verticillata*) is mentioned as a crop in the ancient Chinese *Shijing* (Book of odes), a compilation of poems dating to between the seventh and eleventh centuries BC. Jia Sixie mentions mallow in his famous ancient text on farming practices, *Qimin*

Yaoshu (ca. AD 544).[16] In his *Simin yueling* (Monthly ordinances of the four classes of people), Cui Shi (AD 103–71) also mentions growing mallow as a crop, supporting the idea that it was a common food.

More detail comes from Tao Yuan-ming (AD 365–417), the greatest poet of the Wei period in China, who wrote extensively about his humble farm in the countryside. He claimed that he grew mallow along with fruits and grains on his farm, and that mallow was his favorite vegetable. Tao was the first Chinese author to make reference to the "heating" and "cooling" properties of foods, a concept based on the humoral system laid out by Galen (129–ca. 210 BC) and others, which spread along the early Silk Road from the Hellenistic world.[17] It appears that by the time of Tao, mallows (like the chenopods discussed below) were considered rustic crops, praised for their simple and healthy qualities—they were essentially the soul food or Grandma's home cooking of China in the mid-to-late first millennium BC. Historians studying the Warring States period in China note that ordinary people ate mallows and onions.[18] Some of these historians have suggested that mallow was once very important to the diet of the common people in China but later became stigmatized because of its association with poverty. This shift is reflected in the poetry of the Song period.[19] As the cruciferous vegetables gained popularity, the chenopods and mallows came to be considered poor people's food or famine food.

Although it was probably far less prominent in antiquity than mallow, spinach (*Spinacia oleracea*) may have moved along the southern routes of the Silk Road. Its origins are still debated, but it likely hails from somewhere in southwest Asia. The founding scholars of plant cultivation, Alphonse de Candolle and Nikolai Ivanovich Vavilov, placed the crop's origins in "Persia" and southern Central Asia, respectively.[20] De Candolle concluded from linguistic data that the crop was cultivated in Persia by at least the time of Classical Rome and from there quickly spread across southwest Asia.[21]

Other historians have placed the plant's origins in southwest Asia around the sixth century AD.[22] In China, spinach was known as *pocai*, or Persian greens (derived from *Posi*, the term for people from southwest Asia or Persia).[23] There are no known references to spinach in Europe before the Arab conquests in the eleventh century AD, when it appears to have spread with other vegetable crops, like purple carrots, to Spain. De Candolle cites Ebn Baithar (1235), who quotes earlier textual works in stating that spinach was commonly cultivated at Nineveh and Babylon, but the specific meaning of this passage is uncertain.[24]

The crop may have spread along the southern foothills of the Himalaya and reached Nepal before making its way to the Tang dynastic center at Chang'an.[25] Historical sources suggest that it was introduced to China proper in the seventh century AD. If this the case, the crop may have originally spread with the Hu, or people from Iran.[26] A Tang Dynasty text suggests that the crop was presented at the Chinese imperial palace by Buddhist monks, but no archaeological data thus far support these textual sources. (And no source explains how spinach became the source of the cartoon character Popeye's superhuman strength in twentieth-century America!)

A wide range of wild vegetables was likely collected and consumed by the people of Central Eurasia in the past. What role these potherbs played in the cuisine and culture is uncertain, and archaeobotanical data are scarce because of the perishable nature of the foods and the rarity of preserved seeds. However, historical texts and ethnobotanical accounts of modern herders in Central Asia may hint at long-standing culinary traditions.

Among the hundreds of wild plants that were likely foraged in the past are varieties of *Amaranthus* and *Chenopodium*, sister genera in the Amaranthaceae family. These have been collected from the wild for millennia by people across the entire Northern Hemisphere and parts of the Southern Hemisphere (notably in the Andes). They were also domesticated as vegetable and grain crops in multiple regions. Three species of amaranth were domesticated for use as grain in Mexico and the Andes: *A. hypochondriacus*, *A. cruentus*, and *A. caudatus*. All are grown in China today, but they originated in the Americas. A fourth domesticated species, *Amaranthus tricolor*, hails from East Asia and is sometimes referred to as Chinese spinach. *A. tricolor* was cultivated in parts of China in ancient times, both for its grains and for its leaves and stems. The name "Chinese spinach" can refer to more than one species, and wild *Amaranthus* species are also sometimes clumped under this vernacular. It is generally accepted that *A. tricolor* was originally domesticated in India or southeast Asia.[27] Another species, *A. mangostanus*, was cultivated across northern China by the fifth or sixth century AD.[28]

Like the amaranths, the chenopods have had a long coevolutionary relationship with humanity. One species of grain chenopod (*C. giganteum*) appears to have been domesticated in China and was once cultivated as far west as the Himalaya; it was likely first domesticated in the Longshan culture of eastern China (ca. 2400–1900 BC). The genus is best known in traditional cuisines of the Americas: at least two species were domesticated in South America, *C. pallidicaule* and the now popular *C. quinoa*. Morphological evidence for domestication of *C. quinoa* in specimens recovered from archaeological sites near Lake Titicaca in Bolivia dates back to 1500 BC.[29] *C. berlandieri* ssp. *nuttalliae*, from Mexico, is likely a more recent domesticate. Another more recent *Chenopodium* domestication in Mexico includes a landrace variety that is harvested for its broccoli-like inflorescences, called *huauhtzontle*.[30] *C. berlandieri* ssp. *jonesianum* was independently domesticated in eastern North America as far back as 1800 BC.[31] Still other species have been collected from the wild in both the New and the Old Worlds and eaten as potherbs or harvested for their small grains. Notably, *C. album* has been collected from the wild and possibly maintained in China for millennia.[32] The same species was also collected from the wild across Eurasia and possibly even cultivated in Europe in early prehistory.[33]

In China, chenopods may have been independently domesticated more than once. Chinese texts dating back to the period before the Qin Dynasty, in the first millennium BC, suggest that *Chenopodium* greens were commonly consumed by the general populace. Soupy stews, called *cai geng*, and several distinct kinds of broth are mentioned in the early literature, including *li geng*, or *Chenopodium* broth.[34] Archaeobotanists are still working to determine exactly where and when chenopod plants were first cultivated for their grains, but the use of the plant as a leafy vegetable apparently has a long history.

As with mallow, these greens came to be shunned as the food of the poor in China.[35] Confucius is said to have consumed chenopod greens in a broth as a symbol of rural virtue and simplicity.[36] He is said to have eaten this meal while traveling through China during his impoverished period.[37] During the Yuan Dynasty (1260–1368), Guo Jujing wrote the *Twenty-Four Exemplars*, a compilation of short stories modeling respect for one's parents and family. In the fifth story, the main character, Zhong You, is a rich man longing for his youth, when he lived in poverty and worked hard to take care of his parents. To illustrate how poor they were, Guo Jujing notes that the family ate only vegetables, and he uses a Chinese character that is often translated as *Chenopodium*.

The *Zhuangzi* is one of the two foundational texts of Daoism, dating from the third century BC. It comprises a series of anecdotal stories, one of which features a particularly impoverished character who is said to use a walking stick made out of a dried stalk of *Chenopodium*.[38] While this is clearly an exaggeration, as the stem could not support a man's weight, the motif of the pigweed staff was picked up by later poets, from Du Fu and Han Shan to the Japanese haiku writer Matsuo Basho, to celebrate the virtues of poverty and simplicity.[39]

North of the ancient Chinese capital of Chang'an, the Han Yangling Mausoleum was erected for the fourth emperor of the Western Han Dynasty, the Jing emperor Liu Qi (188–141 BC), and his wife. The tomb comprised eighty-six outer burial pits; a layer of preserved plant material and grains was found at the bottom of pit number 15 (DK15), radiocarbon dated to between 300 and 200 BC.[40] In addition to rice and both broomcorn and foxtail millet, the layer contained what appears to be domesticated *Chenopodium*.[41] These

Western Han grains resemble *C. giganteum*.[42] *Chenopodium* grains, likely cultivated, have been reported from the Haimenkou site in Jianchuan County, Yunnan Province, China.[43]

As with the wild mallow seeds discussed above, carbonized remains of *Chenopodium* seeds have been found in abundance in nearly every archaeobotanical assemblage from Central Eurasia. In many cases, they far outnumber domesticated grains.[44] However, it is not possible to determine whether they are the result of human or animal foraging. *Chenopodium* plants are indicator species of active or abandoned pastoral camps, and they become more dominant in the vegetation around pastoralist sites as a result of herding activities.[45] In many ways, *Chenopodium* plants are the quintessential case study for Edgar Anderson's dump-heap hypothesis for domestication, whereby certain plants are inadvertently brought under human dominion because they flourish in the disturbed soils of midden or trash deposits on the outskirts of human settlements.[46] Furthermore, analyses of dung from ancient sites shows that chenopods were a major component of the herd-animal diet.

Despite the difficulties of determining whether wild *Chenopodium* seeds in the archaeobotanical assemblages came from animal or human food, several scholars have argued that this plant was foraged for human food by early Central Asians.[47] At least one Russian archaeologist has claimed that it was collected by early Iron Age people in the Minusinsk Basin in the Altai Mountains.[48]

In my own archaeobotanical research in Central Asia, I have found abundant remains of *Chenopodium* seeds, especially at Tasbas, Begash, Mukri, and Tuzusai, all in the Semirech'ye region of eastern Kazakhstan, and Tashbulak in Uzbekistan.[49] Undifferentiated Amaranthaceae and Chenopodiaceae seeds were recovered at the site of Donghuishan in Gansu, dated between ca. 1550 and 1450 BC.[50] *Amaranthus album* seeds were recovered at Gashun-Sala in the Caspian steppe.[51] At the Late Shang period site of DGS PI HI, archaeobotanists found *Chenopodium album*.[52] *Chenopodium* seeds were also recovered from Xiongnu-period sites in Mongolia, from Botai culture sites in northern Kazakhstan,[53] and sites in southern Central Asia, such as Adji Kui and Ojakly.[54] This genus is very common in archaeobotanical assemblages from Europe as well.[55]

Excavations were conducted at the long-term settlement of Krasnosamarskoe and the herding camps of Peschanyi Dol 1, 2, 3, and Kibit 1 by the Samara Valley Archaeological Project, with the aim of understanding settlement patterns and herding during the second millennium BC in the heart of the Eurasian steppe.[56] Krasnosamarskoe is one of several large settlements along rivers of the western steppe, in the middle Volga region. There are similar settlements along the Samara and lower Sok Rivers.[57] At these sites, members of the Srubnaya culture established houses with wooden walls and roofs.[58] Extensive archaeobotanical analysis at these sites produced no evidence of domestic crops, but the excavators have pieced together an economic model for this community that incorporates animal herding and foraging of wild plants. They noted in particular the importance of the wild grain *Chenopodium album*.[59] High percentages of *C. album* were found in assemblages at Peschanyi Dol 1, 2, and 3 (2 in particular), as well as at Krasnosamarskoe and Kibit 1 and 2.[60] A number of *Polygonum* nutlets were found in combination with a massive amount of *C. album* in a waterlogged pit (feature 10) at Krasnosamarskoe, which may represent a grain store or cache.[61]

Hans Helbaek, a well-known early participant in the debate over the origins of agriculture, asserted that *Chenopodium* was certainly a major food source for people across Europe in prehistory. "This is proved for the Danish Iron Age by finds in the stomachs of corpses found in bogs and pure deposits of *Chenopodium* and *P[olygonum] lapathifolium* seeds in burnt houses in Jutland, and disproportionate amounts of *P. convolvulus* in food remains and grain deposits in Central Europe and Denmark demonstrate the utilization of these large fruits."[62] In his 1950 analysis of the stomach contents of the Tollund bog body (one of several well-preserved bodies found in northern

European bogs), Helbaek analyzed the seeds in the stomach, noting that the last meal was either a bread or a gruel containing roughly forty different kinds of seeds, including barley, flax, chenopods, and *Polygonum*.[63] Similar "last suppers" have been found in other bog body stomachs. These mixed-grain or muesli gruels were likely a common meal across Eurasia, and prehistoric farmers probably did not distinguish between domestic grains and wild seeds in the same way that we do today.

Elsewhere around the world, *C. album* is noted as a food in Russia, specifically as a famine food.[64] Both *C. album* and *C. murale* were, and are, used throughout southwest Asia as a salad green and potherb. *C. album* was once cultivated as a bread grain in southwest Asia. *C. opulifolium* was used as a potherb in the Mediterranean world and east all the way to Iran.[65]

Historical sources also attest to the collection of chenopod seeds as a supplementary grain source across Russia and Central Asia. These sources from the steppe go as far back as 1092 in texts from Kiev, Ukraine. Medical reports from the 1840s noted that the main meal of many peasants in Russia was chenopod flour.[66] The practice of collecting wild grains is mentioned in accounts from the Volga region during the famines of the early 1930s and from much of Russia after World War II. The grains were ground into flour in hand mills and processed into unleavened breads, often supplemented with barley or rye flour.[67] These breads, which were sold in markets, were called black breads because of the hard black testa of the chenopod seed. Other Russian accounts discuss the collection of *C. album* seeds (and *C. viride*, a member of the *album* complex), which were ground and baked into bread with other wild grains, such as *Polygonum*.[68]

Because of their hard testa and extended dormancy stage, chenopods are hard to bring under cultivation; however, these traits make them especially well adapted to soils disturbed by human activity in areas such as middens, gardens, abandoned livestock pens, and crop fields. The seeds survive passing through ruminant stomachs and can remain dormant until humans abandon a settlement, after which the chenopods become pioneer plants. Because of their apparently indefinite dormancy, they tend to dominate the soil seed bank, making it impossible to weed them out of any garden permanently (a frustrated lifelong personal observation). At some point, early gardeners must have thought, "If you can't beat them, eat them," and turned their chenopod weeds into salads. Intentionally or otherwise, both chenopods and amaranths have been manipulated and cultivated as crop inclusions for millennia.

Wild greens were collected and, in some cases, cultivated throughout history and prehistory. In most cases, these vegetables have faded into obscurity. The only vegetable that was domesticated in the mountains of Central Asia and persists in our kitchens today is rhubarb (*Rheum rhabarbarum*), wild relatives of which grow across the alpine meadows of Inner Asia and are still collected by people in the Pamir and Hindu Kush.

ROOTS AND STEMS

Carrots

Wild carrots, members of the genus *Daucus*, grow across the Northern Hemisphere and were likely dug up and consumed by humans for tens of thousands of years before they were ever cultivated. This, in addition to the fact that they are perishable and are often grown from fragments of the root rather than from seed, makes it nearly impossible to track down their origins through archaeobotanical evidence. But bits and pieces of evidence, mostly from historical sources, suggest that people in Central Asia may have been familiar with a few of the root crops that we know today, including carrots.

Taxonomists have split the domesticated carrot into two clades that likely followed different trajectories toward domestication. The first clade contains anthrocyanin compounds, and the plants often have purple or yellow taproots; the second group includes plants that contain carotene, the more familiar orange pigment. The purple and yellow clade seems to express the greatest diversity in southwest Asia, especially in the region of modern Afghanistan, where scholars believe it originated.[69] European literary sources trace the spread of anthrocyanin carrots out of southwest Asia by the tenth century, following Islam, into Spain by the twelfth century, and reaching northwest Europe by the thirteenth or fourteenth century.[70] The familiar carrot that Bugs and the Easter Bunny carry around may have arisen much later, possibly hailing from lines bred from wild populations in Holland in the seventeenth century.[71]

Historical sources from China date the introduction of anthrocyanin carrots into East Asia, through northerly routes along the Silk Road, to the thirteenth or fourteenth century. From there they spread to the Sichuan Plains and ultimately into southern China.[72] The first historic references to the carrot in China appear in the Yuan Dynasty (1271–1368).[73] During the Yuan Dynasty, the Mongols maintained an extensive communication network, and Silk Road trade flourished. The carrot likely made its way along trade routes during this period, becoming known in China as *hu luobo*, the Iranian radish, with Hu again referring to people from northern Iran or Central Asia.[74] The earliest Chinese carrots were reddish-purple, red, orange, or yellow. The more familiar carotene carrots were a colonial introduction to Chinese cuisine. Carrots appear frequently on the Chinese dinner table, often as the unsung hero, adding color to salads and serving as ornamental garnishes.

Another close relative of the carrot may have come to Europe via the Silk Road early on; *Sium sisarum*, or skirret, is another member of the Apiaceae family, with a large white taproot. It is still consumed in parts of Europe and Asia but is less common in modern produce markets than the carrot. It seems to have originated in East Asia, but, assuming the historical identifications are correct, it is discussed by Pliny the Elder, who claimed it was one of the favorite foods of Emperor Tiberius (42 BC–AD 37).[75] Hence the vegetable must have made its way from China to the Mediterranean before the Roman period.

Turnips

The turnip (*Brassica rapa* ssp. *rapa*) also traveled along the Silk Road and was introduced to northern China before appearing in the south. As with the brassicas, sorting through possible ancient literary references to the turnip is difficult, because the turnip closely resembles the Asian radish or daikon (*Raphanus sativus*), and it is not always clear which of the two crops is being discussed. However, historians have suggested that the turnip spread into East Asia from the west before the sixth century AD.[76] With the help of modern genetic studies, the story of the *Brassica* crops is slowly being pieced together.

Onions

As with many of the plants discussed in this chapter, pinpointing a center of domestication for the genus *Allium* is a herculean task. Any attempt to understand the domestication of onions, garlic, leeks, chives, shallots, and their relatives is hampered by two facts. First, there is little need to domesticate the plant, as it makes a nearly perfect food in its wild state, already containing high levels of the sulfides that provide its distinctive flavors. Second, its propensity for artificial reproduction and feral growth complicates genetic studies. Members of this genus grow across the Northern Hemisphere in nearly all ecological conditions, and many are interfertile. Bulbs are even less likely to survive in the archaeobotanical record than other root vegetables (such as tubers and rhizomes), which are only rarely preserved when carbonized or desiccated. Wild onions resembling our modern domesticated onions grow across Central Asia from the steppe to the mountains. Some historians have suggested that there were three centers (or areas) of domestication: in Central Asia, in China and neighboring regions, and in southwest Asia and the Mediterranean.[77] This seems to me like an oversimplification. Some form of wild onion was undoubtedly part of the earliest European cuisine, and it was clearly a component of the early Central Asian diet and part of the Chinese culinary tradition from the outset. (In some South Asian cultural traditions, however, onions are considered unclean and are not eaten.)

Several nineteenth-century European explorers in northern Central Asia noted that wild *Allium* bulbs (e.g., wild field onions or bear onions, wild garlic, and leeks or ramps) were collected and stored for the winter and that wild onions were sometimes fermented for longer storage.[78] Wild *Allium* species grow in abundance across Central Asia today and can easily be picked or dug up in large quantities. According to legend, the young Temujin (Genghis Khan) and his mother had to survive by collecting wild onions and hunting small birds after they were banished from their group.

Lilies and Other Geophytes

Another bulb plant that provides a food source for humans is the lily. Members of the Liliaceae family have long been valued not only as ornamental plants but also as a source of nutrients, especially in the northern latitudes and higher elevations of North America, Europe, and Asia. People in China and the Ainu people of Japan have traditionally collected several species of wild lily bulbs.[79] The roots and dried flower buds of lilies are mentioned in the *Shijing* as foods.[80] The common tiger lily (*Lilium tigrinum*), popular in flower gardens today, was domesticated in China as a food crop (but should not be confused with the tiger daylily [*Hemerocallis fulva*], which also has an edible root and originates in Asia, and also appears in the *Shijing*).[81] Several other species of lilies may have been cultivated in China, but they have not left a clear archaeobotanical record. Lily bulbs were a highly sought-after food in China and were both imported and exported along exchange routes, especially in the south.

Geophytes (underground storage organs, such as bulbs, rhizomes, roots, and tubers) were important sources of carbohydrates for early peoples in the Altai Mountains, such as Kazakhs and Tuvans, and people farther north, such as the Yakuts. Nineteenth-century European explorers in Central Asia noted the importance of wild plants in the diet, particularly wild roots and tubers.[82] As Sevyan Vainshtein recounts, early explorers traveling around Tuva in the remote mountains of southern-central Russia in the late eighteenth century observed that, starting in the middle of August, the Tuvan herders traveled across the mountains to gather lily bulbs. These explorers saw the seasonal migrations as a means of collecting wild plant foods for human consumption rather than of finding herd-animal forage.[83] In the mid-nineteenth century, V. L. Priklonskii observed the same reliance on foraging among the Yakuts of southern Siberia, as well as other groups of mobile pastoralists in the Altai Mountains, such as the Altai-Kazakhs.[84] Ethnographers studying Central Asian mobile populations have made similar observations.[85] Many of the harvested wild roots were spring ephemerals, such as *Erythronium*, that had to be harvested in late spring or early summer after the plant had restored its root nutrients. *Erythronium* bulbs were dried and stored in large sacks.[86] Fresh bulbs were prepared for eating by putting them directly on the ashes of a fire or cooking them with other foods.

A number of geophytes were harvested in late summer or fall, including those from *Allium* spp., *Lilium* spp., *Paeonia anomala*, *Polygonum viviparum*, *Sanguisorba alpine*, and *S. officialis*.[87] *Lilium* bulb harvesting started in August, and *Sanguisorba alpine* roots were harvested in July and August.[88] Historians, ethnographers, and general observers of Kazakhs and other Central Asian herding populations have also noted that wild and cultivated plants were (and are) particularly important parts of the winter diet, which is often lacking in vitamin C. *Allium* bulbs, which can be stored for the whole winter, are particularly high in vitamin C.[89]

CONCLUSION

While Central Asia has often been described as a pastoralist realm, vegetables were as important in the economy as grains and fruits from at least the second millennium BC. However, because many roots and leafy vegetables leave no evidence in the form of preserved seeds or archaeological remains, it is almost impossible to piece together what roles they played in the human diet. In addition, the traces left by the burning of herd-animal dung often overshadow any evidence of the collection or cultivation of plants such as mallow or chenopods. The piecemeal historical sources and ethnographic records can provide only the faintest idea of the vegetative plants that people were eating in Central Asia before the organized trade routes were formed.

CHAPTER TWELVE

Spices, Oils, and Tea

When you walk through the market bazaars in Almaty, Ashgabat, Bishkek, Bukhara, Kashgar, Tashkent, or Urumqi, your nose leads you automatically to the tables of the spice vendors—urging you past the butchers with their aged, salted, and cured goat flanks, sausages, and internal organs, and the tables of fermented dairy products, such as kumiss and qurt. The colorful mounds of powdered plants and dried leaves, seeds, fruit coats, stems, roots, and flowers are a feast not only for the eyes but also for the nose and tongue. Their scents mingle with all the other pungent aromas of the market to create the unique smell of the Silk Road. But how did all these distant flavors come together? What force brought these spices from the far corners of Asia to the deserts and mountains at the heart of the continent? This is a story that could fill volumes. In this chapter, I trace its broad outlines. I also touch on the stories of a handful of oilseed crops that have enjoyed the spotlight in cuisines of certain parts of Central Asia, and I end with a look at one of the most important crops to move along the Silk Road, a plant that has changed the course of history time and again: tea.

SILK ROAD SPICES: THE TASTE OF ASIA

The culinary magic of the many regional ethnic cuisines of China rests on a mastery of the secondary compounds in plants, which lend spices their distinctive flavors. However, not many of these spices can be traced back more than two millennia in Chinese history. Chinese cuisine began to blossom with the historic opening of the Silk Road during the Han Dynasty, as well as the expansion of the empire into the south and the establishment of trade routes into southeast Asia. The Tang Dynasty's extension of trade routes deep into India and to the islands of the Pacific and Indian Oceans laid the real foundation of what we think of as traditional Chinese food. Here I mention only a few of the key spices of the merchants at the ancient Central Asian bazaars that also flavor the foods enjoyed in Europe and North America.

Many of the spices that are now prominent in southern European and East Asian cuisines originated in southeast Asia and spread along the southerly paths of the Spice Routes to the Mediterranean. For example, ginger originated in tropical forests of southeast Asia, but it was known in southern Europe by the first century AD.[1] The best preserved of the thirteenth-century cookbooks from the Islamic world was written by an anonymous author in Syria (and recently republished in English under the title *Scents and Flavors the Banqueter Favors*). This book mentions a wide range of spices and herbs used in the kitchens of the Ayyubid elites, including agarwood, cinnamon, citron, coriander, fennel, garlic, jasmine (and other scented flowers), marjoram, musk, onions, poppy and sesame seeds, safflower, sandalwood, sugar, and sumac.[2]

Black pepper, from species in the genus *Piper* from tropical South Asia, reached China as early as the Han Dynasty and is mentioned in the *Book of the Han*.[3] This spice, ubiquitous today, became the main cargo of the Spice Routes during the Roman period, and ships regularly landed on the Malabar Coast of southwestern India to pick up dried *Piper nigrum* to transport to Rome. Ancient peppercorns have been recovered from a number of sites around the Roman world, including the Egyptian ports of Berenike and Quseir al-Qadim.[4] As several Classical texts attest (notably the *Periplus Maris Erythraei*, which recounts the treks of merchants out of eastern Egypt), the spice would have come up the Red Sea by boat for transport overland to the Mediterranean. To appease the attacking Visigoths in the early fifth century, Rome paid a ransom that included 5,000 pounds of gold, 30,000 pounds of silver, 4,000 silken tunics, 3,000 hides dyed scarlet, and 3,000 pounds of black pepper.[5] Unfortunately for the Romans, this just whetted the Visigoths' appetites for black pepper and silk. At Quseir al-Qadim, a significant number of the seeds were also recovered from contexts dating into the Islamic period.

Pepper continued to be the spice that launched a thousand ships long after the siege of Rome. By the twelfth century, Venetian and Genoan merchants dominated the spice trade in the Mediterranean, and competition between the two cities ultimately led to the wars that put Marco Polo in prison, where he supposedly recounted the tales of his trek to China. In the 1490s, Europe's hunger for pepper and other spices—such as star anise, turmeric (*Curcuma domestica*), cardamom, cloves, nutmeg and mace, and cinnamon—led Christopher Columbus, Vasco da Gama, and other enterprising adventurers to set sail for distant lands and unexplored seas. Cloves became one of the most important commodities of the nautical spice trade during the colonial period; they were not carried on the early routes of the Silk Road, as the entire supply came from a few islands in the Maluku (Moluccas) archipelago.[6]

Until the maritime spice routes were established, however, spices traveled overland. Similarities in the root words for the names of star anise in Mandarin, Cantonese, Farsi, Urdu, Macedonian, Spanish, Russian, Latvian, French, and German attest to its distribution by Farsi-speaking Sogdian and Persian merchants along the Silk Road.[7] Star anise is a tropical spice, originating in southwestern China and Vietnam, although it is now cultivated across southeast Asia.

There were four indispensable spices in Classical Roman cuisine: coriander, cumin (*Cuminum cyminum*), dill (*Anethum graveolens*), and black cumin

(*Nigella sativa*). They have been found preserved at sites in the far corners of the empire, from Egypt, the shores of the Red Sea, and across the Mediterranean.[8] While all four of these herbs were well known on the ancient Silk Road, cumin and black cumin were two of the most prominent flavors in Central Asia in the past and remain so today. Coriander is an annual that is harvested for both its dried seeds and its green leaves; the leaf form of the herb is often called cilantro. It grows wild in parts of the southern Mediterranean and is a staple of the cuisines of southwest Asia and the eastern Mediterranean. It can be traced back to ca. 6000 BC at Nahal Hemar Cave, near the Dead Sea, just northwest of Mount Selom in Israel; and at Atlit-Yam, near the coastal town of Atlit, also in Israel.[9] Preserved coriander seeds have been recovered from many sites in southwest Asia, including Turkey and Syria; the earliest finds date to around the fifth millennium BC.[10]

The first mentions of coriander in literary sources from East Asia come from the sixth century AD and a text from the eighth century.[11] In early Buddhist tradition, the spice was one of the five fragrant vegetables that were forbidden to monks and geomancers. In the Judeo-Christian tradition, coriander seeds are mentioned in Exodus, Numbers, and the Talmud. The herb is also mentioned by Aristophanes, Theophrastus, Hippocrates, Dioscorides, Pliny the Elder, and Colomella.[12] Mentions of the spice in an ancient Zoroastrian text called the Bundahishn have led some historians to argue that coriander cultivation was prominent in ancient Persia as well.[13] Literary sources from China show that coriander was present there by the sixth century AD, and it was probably a common item of trade on the Silk Road.

Asafedita (*Ferula asafoetida*) is another important spice of Central Asian origin, which some scholars think is related to the legendary *silphium* spice of ancient Rome. It is a wild plant that grows in hyperarid regions of Iran and Afghanistan, as far north as Fergana. The plant excretes a highly pungent resin that dries hard and is crushed into a spice. It grows wild in the

Kyzylkum Desert along the edges of key stretches of the Silk Road through the Zerafshan region.

Saffron, an important spice in Persian and later in Arabic cuisine, was also a prized commodity along the Silk Road. It commanded such a high price and was so light that it was an even more profitable trade item than silk. It is still the most expensive spice in the world. The dried stigmata of a type of crocus flower (*Crocus sativus*), saffron provides a brilliant yellow hue to otherwise dull-colored dishes, especially rice. The flavor is also strong and distinctive. One kilogram of the spice requires laborious handpicking of the 3-centimeter-long stigmata from approximately 150,000 flowers, or about two acres of cultivated fields.[14] It is generally accepted that the modern domesticated variety of saffron crocus, which is the most widely harvested of all the *Crocus* species today, originated from genome duplication as a result of hybridization between two of the several wild species that were harvested in the past. One of the parent species of this polyploid variety was most likely *C. cartwrightianus*, which currently grows across most of Greece; the other may have been *C. thomasii* (or a close ancient relative), which is also distributed in the Mediterranean region. Saffron prefers hot, dry summers and mild winters. Because the flowers are so delicate, rain or frost during blooming can destroy an entire crop.

Literary sources attest that the flower was grown in Vedic India (1500–500 BC). A wall at the palace of Minos on Crete bears a three-thousand-year-old depiction of a crocus flower. It may also be represented on Minoan pottery and frescos from ca. 1645 BC, notably in an ancient fresco titled *The Gathering of the Crocus from Akrotiri*, in Santorini.[15] Thus saffron was probably widely harvested in antiquity. The spice likely reached China via Kashmir, but it would have been harvested in Persia, because the crop would have been unsuccessful on the northern routes of the Silk Road. Abul Fazl, the vizier of the Mughal emperor Akbar, notes that saffron was a major crop near the village of Pampur in Kashmir in the early 1600s, stating that "the feast of cups was held in a saffron field. Groves on groves, and plains on plains were in bloom. The breeze in that place scented one's brain. The stem is attached close to the ground. The flower has four petals, and its colour is that of a violet. It is of the size of a champa flower, and from the middle of it three stigmas of saffron grow."[16] An old Chinese name for saffron, meaning "Tibetan safflower," supports the idea that it reached China through the southerly passes of the Pamir or Himalaya range.[17]

In India, saffron is cut with safflower (*Carthamus tinctorius*), which produces a similar deep yellow-orange pigment. In literary accounts safflower is often confused with turmeric, which comes from the dried tubers of the *Curcuma* genus, a relative of ginger. Safflower originated in southwest Asia, where close wild relatives still grow, and it has traditionally been cultivated in southern Central Asia, Afghanistan, and Iran. The oldest evidence for its cultivation comes from Egypt in the middle of the second millennium BC.[18] Safflower made its way to China, where it was used as a dye as well as a spice, by the third or fourth century AD. It is mentioned in an early Chinese text under a different name; there are clearly some discrepancies in these translations.[19] A recent synthesis of archaeobotanical finds of safflower seeds shows that it first appeared in Syria around 3000 BC and from there spread to Turkey, the Balkans, southeastern Europe, and Egypt.[20]

Another interesting group of flavorful plant products that moved along the ancient exchange routes came from the inner bark of certain woody plants, notably members of the cinnamon clade. While many species of wild cinnamon have been used for their fragrant secondary compounds, one of the most famous species in antiquity was *Cinnamomum cassia*, called Chinese cinnamon, or simply cassia. This spice has a flavor at least as powerful as its more familiar relative, the common cinnamon (*Cinnamomum verum*). Like cinnamon, it comes from the inner bark of a tree that has high concentrations of essential oils and cinnamaldehyde, providing the familiar spicy flavors. The tree grows across southeast Asia and is cultivated in Guangdong and Guangxi Provinces of southern China. In 216 BC, the first emperor of the Qin Dynasty, Qin Shihuangdi, conquered a region famous for cassia cultivation and named it Kweilin, after the ancient Chinese name for cassia, *kwei-shi*.[21] Now known as Guilin, the city was the former capital of the province during the Ming and Qing Dynasties and is still one of the largest cities in Guangxi Province.

Cassia was carried by Sogdian and Persian traders along the Silk Road long before common cinnamon was known in Central Asia. In Persian and Arabic, cinnamon was referred to as *dar-sini* or *darsini-sini*, names that translate as "fragrant wood of China" or "Chinese cinnamon from China," emphasizing the fact that much of the cassia in Persia traveled the southerly routes of the Silk Road or Spice Routes from China.[22] The Uighur people still use the word *dar* as a generic term for spices, showing the historical importance of cassia in parts of Central Asia.

Herodotus describes cassia as one of the main spices of Arabia, along with frankincense (*Boswellia* sp.), myrrh (*Commiphora* sp.), cinnamon, and ladanum (*Cistus creticus*, a kind of resin).[23] While he was aware that they arrived at Greek ports with Arab merchants and that they fetched a high price in the market, he recounts only fanciful tales about how they were obtained and the plants they were derived from. For example, he insists that frankincense grew on a plant guarded by dragons or winged serpents and was obtained at great peril by the people of the Arabian Peninsula. Cassia, he claims, grew in a deep lake whose shores were guarded by giant bats. Cinnamon came from the twigs used by giant birds to build their nests on high cliff faces.

Herodotus was not the only Classical author to spread these tales; Theophrastus also provides fanciful stories of the origins of spices. While he clearly did not know what plant cassia came from, he was at least aware that it was brought via Arabia by traders, along with myrrh and frankincense. The more skeptical Pliny claims that these tales were invented by Arab traders to raise the prices of the spices. These stories show how displaced the consumer was from the production system and the products in the globalized market economy of the Spice Routes.

Arguably the most distinctive flavor of the Spice Routes is that of the two-seeded berries from the *Zanthoxylum* tree, the prickly ash or Sichuan peppercorn tree. This uniquely piquant spice, the key ingredient in many traditional Sichuan dishes, burns while simultaneously numbing the tongue. It is central to the cuisine and identity of the people of the Sichuan Plains and the Himalaya, but it was not as readily adopted outside Inner Asia. It was the only hot spice known in central China before the postcolonial introduction of the chili pepper from the Americas. Today, however, it is often absent from Sichuan dishes served in the United States. It was banned as an import in the late 1960s because it was discovered that the fruits carry a bacterium that causes citrus canker, a threat to the orange orchards of the southeastern states.[24] In 2005, the importation of heat-treated, nongerminating Sichuan peppercorns was permitted, and they have since reappeared in Asian specialty markets across North America.

This spice has a long history in China, especially in the western regions. The berries are mentioned several times in the *Shijing*. The early texts seem to mention both *Z. piperitum* and *Z. simulans*, and possibly other species too. Of several species of the tree that have been harvested for their berries in the past, *Z. piperitum* is the most widespread (found in the wild across China and in Japan) and the most widely cultivated today. Other species would have been traded along the Silk Road, notably those endemic to the Himalaya mountains. A few ancient seeds, possibly *Z. piperitum*, were recovered from the Islamic trading port of Quseir al-Qadim in Egypt (AD 1040–1160).[25]

OILSEED PLANTS

In today's Central Asian cuisine, almost every dish is cooked in sheep fat, often from the tail of an Afghan breed of fat-tailed sheep. However, oils derived from plants were clearly known in the region and are still important in some regional cuisines, such as that of Uzbekistan. We know very little about how oilseed crops were processed or pressed in the past, but traditional pressing practices are still used in many areas of western China, especially to produce rapeseed oil (*Brassica napus*). A familiar sight in Europe in the early summer, bright yellow fields of blooming rape plants also stretch across the Sichuan Plains of central China. Entering any small village in the winter, a visitor immediately encounters a unique and strong woody, burning scent—the smell of rapeseed oil being boiled down. Villagers process the oils slowly throughout the winter, because the slow heat required for the process also keeps their houses warm. A similar scent may have been familiar in many Central Asian villages before the introduction of processed oils and the heavy reliance on animal fats for cooking. I have recovered archaeobotanical remains of flax (*Linum usitatissium*) and lallemantia (*Lallemantia iberica*) in southern Central Asia dating to the second millennium BC. Other oilseed crops, including hemp, cotton (*Gossypium arboretum* and *G. herbaceum*), poppy (*Papaver somniferum*), and, later, gold of pleasure (*Camelina sativa*) and sesame (*Sesamum indicum*), spread across the Iranian Plateau to the Indus or vice versa.

Hemp and cotton, like flax, are oilseed crops that have additional uses as fiber crops. The history of their origin and spread is complicated and still contains significant gaps, but we know that both crops were already fully domesticated by the start of the second millennium BC. Hemp likely originated in China; at least one species of cotton came from India and another from elsewhere in South Asia.

Sesame

The small, flat white or black seeds that you often find on your bagel or hamburger roll are the oily seeds of the sesame plant. This crop has a long and interesting history, mostly in South Asia. Many historians and scholars have claimed that it spread westward into Central Asia from India, but the story of its journey along the Silk Road is not really clear and has almost no scientific backing. Historians also like to claim that sesame moved into China from Central Asia during the Han Dynasty.[26] According to legend, the great general Zhang Qian brought it back with him to the Han court. Again, however, there is no archaeobotanical evidence to support that account.

The crop likely originated in India, from *S. orientale* var. *malabaricum*. The oldest archaeobotanical evidence for the seeds comes from Harappa proper in the Indus Valley, roughly 2600–2000 BC.[27] There is no evidence for the crop in early Central Asia, and it is not clear when it reached China. Further archaeobotanical investigation in Central Asia may help clarify the sesame story.

Preserved sesame seeds are occasionally found in later medieval settlements across southwest Asia, but rarely in high abundance. One exception is a concentration of the oily seeds in a hearth from the mid-eighth to the ninth century from the site of Tell Shheill I in Syria. Seeds have been recovered in low abundance from sites throughout the upper Mesopotamian region in Syria dating to between the eight and thirteenth centuries.[28] Sesame clearly played some role as a summer crop and is mentioned in written sources; however, its low prevalence in archaeological sites leads one to wonder how prominent it really was in crop-rotation cycles and whether it ever spread into regions such as Central Asia. The cultivation of sesame is often glossed over by early Arab geographers. For example, in his eleventh-century agricultural calendar, Ibn Wahshiyya discusses the cultivation of many crops, such as rice, in great detail but mentions sesame only in passing.[29]

Flax

Flax appears to have been the world's first oilseed crop; it is part of the southwest Asian founder-crop complex and was brought under human dominion at least ten thousand years ago in the Fertile Crescent.[30] Domesticated flax spread across western Eurasia during the Neolithic. Linen, which is spun from the fibrous stems of the flax plant, was probably the dominant textile source across Eurasia before wool.[31] Because flax is a thirsty crop, requiring over 750 millimeters of annual rainfall or irrigation, it cannot easily be grown in many of the northern regions of Central Asia, and it is likely that preserved fragments of early textiles found along the Silk Road were actually produced elsewhere.[32] Linen remains from Central Asia date back to the second millennium BC. Three seeds identified as "*Linum* sp." were found in period I, level 2 at the Shortughai site in Afghanistan (late third or early second millennium BC), and impressions of *Linum usitatissimum* seeds were found in mud bricks at the site.[33] Linen seeds have been found in Bronze Age levels at Miri Qalat,[34] at Pirak,[35] and across the Harappan world.[36] In addition, I recovered a single seed fragment, which appears to be *Linum usitatissmum*, mixed into a cache of domesticated grains at site 1211 in Turkmenistan (1400 BC).[37] It is not clear whether these early finds of flax seeds represent plants cultivated for grain, oil, or fiber. In northern Central Asia, where flax would be less likely to thrive, wool may have been favored and linen lost its significance during the Iron Age.

Domesticated flax likely spread into China along the southern rim of the Himalaya. There is good evidence for flax in the Himalayas in the cemeteries of Mebrak and Phudzeling in the Jhong valley in Upper Mustang, Nepal (1000 BC to AD 100). The assemblages at these sites illustrate that crops from southwest Asia were spreading into the region and being adopted by farmers along with crops of East Asian origin. Domesticated flax, naked and hulled varieties of barley, bread wheat, broomcorn millet, peas, lentils, and hemp were being cultivated by people at the site by phase 2, ca. 400 BC.[38] Following the same trajectory, flax is among the crops found at Ghalegay, Bir-kot-Ghundai, and Loebanr in the Swat region of Pakistan, dating to 1900 BC or earlier.[39] The sites of Burzahom, Gufkral, and Semthan, all in Kashmir (ca. 2800–2300 BC), have provided archaeobotanical remains of other southwest Asian crops as well, illustrating that flax moved into the region as part of a package of crops. Slightly less straightforward evidence for flax in

Figure 29. Traditional flax harvesting for linseed oil in Inner Mongolia, China, 2010. The farmers are threshing the grain loose from its husk after beating it with brooms. It will be hand-sieved in baskets before the seeds are processed into oil using a stone roller and slowly boiled to concentrate the oils. Photo by the author.

eastern Tibet comes from the high-elevation site of Ashaonao. Because the specimens found at the site are reported to be smaller than domesticated varieties in Asia and wild forms in China, the archaeobotanists on the project suggest that they may be seeds from a wild plant. However, the researchers' nomenclature is somewhat confusing: they refer to the wild flax seeds by various taxonomic classifications, including *Linum ussisitum*.[40]

References to flax in early Chinese texts pose challenges of interpretation.[41] The early name for the plant was *hu ma*, meaning "hemp of the Hu people" from southern Central Asia and Iran. The term is used for both flax and sesame.[42] Both are oil crops, and both probably spread into dynastic China during the first few centuries of the common era, along trade routes through either the southern Himalayan rim or Central Asia. However, while flax originally came from southwest Asia, sesame was domesticated in South Asia, likely India, at least three thousand years ago.

Lallemantia

Lallemantia iberica (dragon's head) is a largely forgotten crop today, but it has been cultivated since antiquity across southern Central Asia, southwest Asia, and southeastern Europe to provide oil for cooking, lighting, varnish, and tanning.[43] Five wild species of the plant grow across southern Central Asia and southwest Asia.[44] Archaeobotanical specimens of *Lallemantia* excavated from sites in the Macedonian region of northern Greece are morphologically similar to *L. peltata*, *L. iberica*, and *L. canescens*, although *L. canescens* has a slightly larger seed.[45] Remains of *Lallemantia* seeds that I recovered at the ancient city of Adji Kui in the Murghab region of southern Turkmenistan also seem morphologically similar to those recovered in Greece.[46] The likelihood that the Adji Kui seeds were cultivated and used as food is supported by the fact that they were found among other cultivated crop seeds, such as wheat, barley, legumes, and broomcorn millet, but because wild *Lallemantia* relatives grow in the region, this supposition cannot be verified at present.

Lallemantia seeds have been noted at the second-millennium BC sites of Mandalo, Archondiko, and Assiros in the Macedonian region, as well as the slightly later sites of Ayios Mamas and Kastanas.[47] The plant may have spread to the region with Central Asian pastoralists, in concert with broomcorn millet and possibly tin alloys.[48] This hypothesis would support the idea that millet reached Europe by a route through the southern Caspian Sea region to Anatolia and eventually to the Balkans. This route of spread may explain why broomcorn millet did not reach the sedentary agricultural centers across southwest Asia until rather late; it was maintained as a complement to a pastoral economy among peoples along the boundary of southern Central Asia and southwest Asia. *Lallemantia* would have fit well into the economy of these transient populations because of its drought tolerance and short growing season, both traits that also made millet favored among mobile peoples in Eurasia.

A CUP OF TEA OR A BRICK OF CAMEL SWEAT

No discussion of plants on the Silk Road would be complete without the inclusion of tea (*Camellia sinensis*). Perhaps no other plant represents globalization as well as the humble *C. sinensis* bush, which has given rise to the traditions of afternoon tea in the British Isles, the dark, rich tea of Assam in India, the sugary and milky tea mixture served across the Russian world, and the spicy chai from South Asia. Its rise to global prominence is linked to the routes of the ancient Silk Road. Japanese legend claims that tea came from the eyelids of a traveling monk, who arrived in China in AD 519 after a long trip from India. Wishing to pray but unable to stay awake, he cut his eyelids off to keep his eyes open. The tea plants sprung from his severed eyelids. A parallel but slightly less gory version of the story exists in China: the Indian prince Bodhidharma, who is often depicted in Chinese art as an overweight, scowling man with bushy eyebrows, clothed in a silken robe, vowed to meditate in a cave unceasingly for nine years. During his meditation, he fell asleep once. To punish himself for breaking his vow, he shaved off his eyebrows and threw them to the ground. The next day, he found a tea bush where his bushy brows had fallen, and after consuming the leaves he was revitalized and ready to resume his meditation.[49]

Given the vast range of true teas in the world—including Dongding oolong, gunpowder, smoked teas, *biluochun*, Earl Grey, Japanese *gyokuro* and *matcha*, Ceylon, the white-tipped Java tea, orange pekoe, pu'erh, *kukicha*, jasmine, *longding*, and *ming ding*—it may be surprising to learn that they all come from one plant. The differences lie in the way the leaves are processed after harvesting. The four main classifications of tea—white, green, oolong, and black—reflect the degree of oxidation, the aging process that the leaves go through before drying, in which they start to turn brown and develop distinctive flavors. Once they are aged according to preference, they are dried for storage. Tsiologists have studied hundreds of varieties and processing methods, but in the end, all these subtle and complex flavors derive from one emerald leaf.

Today *C. sinensis* is grown in many tropical and subtropical regions of the world. The evergreen foliage is not highly tolerant of frost. The bush is visually unremarkable, with the exception of its white or yellow flowers. Today, there are hundreds of landrace varieties cultivated for sale in markets in all corners of the globe. In southern China, some of the highest-quality teas are grown in the mountain foothills, between 1,000 and 1,500 meters above sea level. Most of these Chinese tea shrubs have small leaves. The leaves or leaf buds are plucked from the plant at different stages of development, depending on the type of tea being produced.

People of the Ming Dynasty preferred green tea, but they also perfected the art of oxidizing teas, from the semioxidized oolong to the fully oxidized black tea. This practice arose in the sixteenth century, when Buddhist monks in the Wuyi Mountains started beating and sun-wilting their tea leaves, creating the first black dragon or oolong teas.[50] In Central Asia, Mongolia, and Tibet, tea leaves were oxidized, dried, and compressed into hard bricks from which chunks could be broken off and immersed in water. These brick teas became staples of mobile pastoralists across Central and northern Asia. Tibetans became accustomed to adding yak butter and barley flour, as well as salt, to their tea. In the heart of the tea-growing regions in Yunnan, tea was often prepared with a mix of herbs and flowers, many of which had traveled great distances. Yunnanese teas prepared by the tea-leaf pickers in ethnic groups such as the Bai, Dai, Hani, Miao, Naxi, Tibetans, and Yi were often of a big-leafed variety (*Camellia sinensis* var. *assamica*) and flavored with ground black pepper, salt, and other spices (including chili peppers, when they reached China in the colonial era). The Wei people of southern

China have historically used tea leaves to make sacrifices to their ancestors.[51] The Wei and Yi peoples of Yunnan were also famous for their rich, wood-smoked teas.[52]

The earliest clear literary reference to tea as a beverage comes from a text called "The Contract for a Youth," a fable written by Wang Bao in 59 BC. The youth is ordered to buy tea at the market in Wuyang, a town near Chengdu, and boil it. Tea seems to have been loosely associated with Chinese identity by the time of the Han Dynasty, although the main beverages of that period were still forms of rice beer.[53]

Tea leaves were still collected from wild stands and from cultivated groves simultaneously well into the sixth century. During the Tang period, tea was often dried and ground, and then boiled with salt in iron pots or kettles and sipped from drinking bowls. Tea-drinking customs changed in the Song period, when water was often heated in a separate vessel and then poured over the leaves: for the elites this vessel was a porcelain ewer with a slender spout. Cakes of tea from Yunnan were pounded in a silk sack and then ground; the ground tea was placed directly in the drinking bowl, hot water was poured over it, and then the mixture was frothed with a bamboo whisk.[54]

By the time of the Tang Dynasty, tea had clearly become an integral part of Chinese identity, as reflected in Lu Yu's eighth-century *The Classic of Tea*.[55] The drink spread rapidly across all of East Asia during the Tang and Song Dynasties. By the end of the Song, tea no longer was a drink only for the elites but was considered a necessity for even the most impoverished peasants.

One of the difficulties of tracing the history of tea in China is the vernacular use of the term. As in English, the word used for tea in early China through the Tang period could also refer to infusions made from herbs or other leaves. The same problem arises with alcoholic drinks: some Chinese authors referred to all of them as wine because the term sounded more romantic than beer or grog. However, before the Wei and Tang periods, most of the alcohol consumed in China was beer, and before the Tang and Song periods, tea was prepared primarily from herbal leaves.

Tea features in countless Chinese stories and poems from the past 1,500 years, possibly more often than wine or other alcoholic drinks. One of the most noteworthy of these texts comes from the routes of the northern Silk Road. "The Debate between Tea and Beer," from the Dunhuang Buddhist Caves, appears, in whole or in part, in six different manuscripts from the collection of scrolls and wood plates recovered at Mogaoku in Gansu. The text, which seems to date to the end of the tenth century, is often attributed to an author named Wang Fu.[56] In the story, the personifications of the two beverages quarrel over which of them has the higher merit; in the end, the duel is quelled abruptly when water speaks up.

A famous Chinese poem about tea by Du Yu (AD 222–84), titled "Rhapsody on Tea," romantically describes the brewing process, noting that the bubbles floating to the surface sparkle like snow. One of the best known of the ancient Chinese texts on tea is Lu Yu's detailed study of tea culture across China, *The Classic of Tea*. The three volumes were written between AD 758 and 775.[57] By the Song Dynasty, tea had become the beverage of scholars and poets who wished to free their minds of the influence of alcohol, as well as Buddhist monks who wanted to prepare themselves for meditation and enlightenment.

Some of the earliest archaeological evidence of tea culture in China comes from a complex of eighty-six tombs surrounding the Han Yangling Mausoleum along the Weihe River near Xi'an, excavated by the Shaanxi Provincial Institute of Archaeology between 1998 and 2005. The tomb complex was built for the Jing emperor Liu Qi (188–41 BC), the fourth emperor of the Western Han Dynasty. The excavation of tomb 15 (DK15) recovered well-preserved botanical remains from the bottom of the tomb, and a recent reevaluation of the leafy materials led to the identification of tea leaves. Nearly three thousand kilometers to the west, on the Sutlej

River in Ngari District, Tibet, at an elevation of 4,290 meters, lies the Guryam cemetery, with tombs belonging to people of the Zhang Zhung Kingdom of the second to third century AD. Excavations in 2012 of a burial of a prince or king in the cemetery recovered numerous luxurious burial goods, including silk fragments and preserved tea leaves. Based on phytolith evidence, the scientists analyzing the plant remains concluded that the tea was mixed with ground and roasted barley groats (*tsampa*), similar to the Tibetan barley butter tea that is drunk today.[58] The finds of tea at this high elevation suggest that Chinese merchants were crossing the arduous passes of the eastern plateau from the central plains of China by the second century AD to supply the central routes of the Silk Road with tea, silk, and other items.

Tea has a particular connection to one route of the Silk Road that runs along the southern Himalaya, through Yunnan. Sometimes known as the southern Silk Road, the route is more frequently referred to as the Tea Horse Road. Archaeobotanical evidence from Kashmir and the Swat Valley of Pakistan suggests that it may date to the late second millennium BC.[59]

Branches of this route of the Silk Road started along the Hengduan Mountains, passing along northern Yunnan, not far from the Sichuan border, through Dali and Lijiang, and along the Yuliang Mountains (figure 30). It then passed farther north along the Cangshan Mountains through the town of Zhongdian (now called Shangrila) and through the outpost towns of Deqin, Mangkang, Zuogong, Bangda, reaching Changdu (or Qamdo; not to be confused with Chengdu), high on the Himalayan Plateau, and then continuing to Lhasa or one of the many other tea-trading towns. Some of the routes passed through northern Myanmar and along the border with India or through Nepal, and another route originated in Ya'an, closer to Chengdu on the Sichuan Plains. After reaching northwestern Yunnan, the route crossed a series of passes in the Nyenchen Tanglha Mountains. In these more rugged

Figure 30. One of the paths of the Tea Horse Road: the view through Leaping Tiger Gorge along the Jinsha River in northern Yunnan, near Yulong Snow Mountain and north of Lijiang, 2011. Photo by the author.

passes, especially in the regions of the Tibetan Khampas, where raiding was common, large caravans would team up, often with hundreds of mules all decorated with bells and ribbons to alert people to their arrival.

During the early seventh century, the unification of Tibet and the formation of a vast empire that stretched into northern Yunnan fueled the trade along this route of exchange. When the Tibetans, under the leadership of Songtsen Gampo (AD 604–50), conquered Lijiang and Dali, they solidified their control over the route.[60] Through a marriage alliance, the Tang Dynasty and the Tibetan Empire established a peace that enabled trade and cooperation between the two powerful realms. Legend has it that the Chinese princess Wencheng arrived in the Tibetan capital with silkworms and a Buddhist statue as part of her dowry, along with yak-loads of cakes of tea. The sinicization of the Tibetan elites continued over the next several centuries, with the side effect of the introduction of tea to Tibet.

Tea porters took their merchandise to a number of market towns on the plateau, along with other products such as salt and spices. But most of the goods that Tibetan herders could offer in exchange were far heavier than tea, notably yak butter and dried yak meat; so by the time of the Song Dynasty, it became customary to trade tea for horses, and eventually this pattern of exchange gave the route its name.

Many of the greatest warhorses used by the Song cavalry came from Naqu in Tibet. The Song needed good horses to wage continual military advances against neighboring groups to the north, and so in 1074 they formed the *chamasi*, or Tea Horse Office, to regulate the trade and create new markets.[61] The agency also forced the farmers of eastern Sichuan to sell their tea to the government at low rates; the government officials then traded the tea to the Tibetans in exchange for warhorses. In 1078, one good Tibetan steed would fetch 100 catties (about 110 pounds) of tea, or 25,000–30,000 copper coins.[62] Trade along the Tea Horse Road was disrupted during the Mongol invasions; it was at its strongest during the Ming and Qing Dynasties (1369–1911).

By the seventh century, the elite of Tibet had developed an unquenchable thirst for tea, which spawned a massive industry of farmers, landowners, merchants, and transporters. By the end of the first millennium tea was a staple of Tibetan culture. Chinese control of the tea supply prevented the Tibetans from making further military incursions into the Sichuan lowlands, because the Tibetans did not want to disrupt the flow of tea into the mountains: some scholars have described the situation as "peace through caffeine addiction."[63]

One obstacle that had to be overcome in order to establish a robust tea trade with Tibet was preserving the tea for transport. The journey could be as long as 2,500 kilometers, over permanently snow-covered mountain passes at altitudes of more than 5,000 meters. The tea was carried by heavily laden human porters, horses, mules, and yaks. The fresh green teas to which the Chinese were then accustomed did not hold up well to the rigors of this journey. Teas made from fermented and oxidized leaves, known as pu'erh teas, were more suitable for transport to Tibet. They were usually produced from a broad-leaf variety of the tea shrub that grows in the mountain foothills of western China.[64] These oxidized teas produced a darker, earthier beverage that appealed to the Tibetan palate. The leaves were exposed to extreme cold as well as hot and humid temperatures in the lowlands, and all the time they were jostled on the backs of sweaty horses and mules. Exposure to these conditions, along with the aging that took place during the weeks-long journey, intensified the flavor of the tea, which became a highly sought-after commodity among the elites. It became known to Tibetans as *jia kamo*, or strong bitter tea.[65]

By the time of the Tang Dynasty, the Tea Horse Road was already a well-established highway with a number of key trading posts, such as Shanxi, Menglian, Menghai, and Simao; the Chinese referred to the route as *cha ma dao*, the Tea Horse Road, but the Tibetans simply called it *gyalam*, or the Wide Road, in recognition of its role as the main artery for supplies and communication between Tibet and the outside world. In addition to enabling trade between China and Tibet, the Tea Horse Road carried a steady stream of products into India. During the ninth century, trade on the route was protected and regulated by the Nanzhao Kingdom, with its capital in the Yunnan city of Dali.[66] The Nanzhao army extended its control of commercial interests north into Sichuan and west into Myanmar (Burma) while also maintaining a strong alliance with the Tibetan Empire. After the Nanzhao, the Dali Kingdom retained control over the trade routes until the invading Mongol forces, led by Kublai Khan, disrupted all commerce on the Silk Road in the thirteenth century.

For most of the past millennium in China, tea has been pressed into bricks for drying and transport. These bricks have traditionally been pressed into various shapes, mostly disks of varying dimensions, which often look like mushroom tops. Some had a hole in the middle, like a large Ming Dynasty coin, which allowed the disks to be strung together for transport and carried on the backs of Bactrian camels along the southeastern corner of the Silk Road, or by mules and yaks on the Tea Horse Road.[67] Brick tea was the dominant form of tea in China until loose-leaf tea became popular again during the Ming Dynasty. Today the tea markets in Yunnan, notably in Kunming, still sell brick teas.

Eventually brick tea made its way into Central Asia and along the main routes of the Silk Road. In Central Asia, it was often called *tuocha*, after the Tuo River at the start of the Tea Road. The sweat and breath of the camels, like the sweat of the horses, yaks, and mules on the road to Tibet, purportedly added a unique aroma to the tea, which became known as camel's-breath tea. With the rise in popularity of loose-leaf tea in the Ming Dynasty, camel's-breath tea faded into history. While *tuocha* forms of pu'erh tea are still found in markets in Kunming and some other parts of China today, they appeal to only a small niche market of culinary historians and fanatics (such as myself). The modern brick teas, although strongly flavored, have neither a hole in the middle for stringing nor the the pungency of camel sweat.

Although the intentional aging and oxidizing of tea began centuries earlier to enable its transport to Tibet, in the sixteenth century the Chinese perfected the process. In Anhua County of Hunan Province, farmers steamed and rolled green tea leaves as they had always done, but then they began letting the tea oxidize in hot, humid rooms for several months, during which microorganisms began breaking down the leaves.[68] This process mimicked the aging of the pu'erh teas on the Tea Horse Road, producing a musty aroma and an amber or dark brown hue. Once merchants realized that strongly flavored tea could be produced inexpensively by this method, it spread rapidly.

In the late seventeenth century, the growing demand for tea in Russia led to a new and even more impressive chapter in the history of Asia's tea roads. Traders increasingly took alternative Silk Road routes, traveling farther north and west, to bring tea to Moscow. During World War II, the Tea Horse Road served as a supply route to Tibet, and the British and American armies formulated plans to send supplies over the mountains to support China in its conflict with Japan. Thereafter, its importance slowly diminished, and the romantic, dusty paths through the mountains have been paved over and lined with stores and factories.

The Silk Route was not the only export channel for Chinese tea. Before the fall of the Southern Song Dynasty in 1279, merchants turned their attention toward nautical routes, and the port of Zayton in southeastern China became one of the busiest trading hubs in the world. Sea transport altered the range of products moving out of China. The ships out of Zayton not only introduced the world to tea but also disseminated the cultural objects associated with tea consumption in China, including porcelain teapots and cups.[69] Too heavy and fragile to move along the Silk Road, porcelain became a luxury export to Europe. The teapot is an invention of the early period of the Ming Dynasty (1368–1644). At around the same time the first true porcelains were fired, an art that was perfected in the 1400s.

Few people realize that the blue and white teacups and plates in their grandmother's china cabinet are a result of the Mongolian conquests. These trace their origins back to the Yuan Dynasty, when Iranian cobalt technology was imported and used to create blue motifs on Chinese porcelains. These dishes were originally created for the great feasts of the Mongolian court. In traditional Mongolian Tenggri shamanism and Buddhism, sky blue is a sacred color: hence, the feasting dishes fashioned during the Yuan Dynasty were made of white and blue porcelain. Later, similar wares would be stacked in crates onto seafaring vessels making their way to Europe to touch the lips of royalty from Portugal to Russia.

The popularity of tea eventually extended west from Tibet to Islamic regions of Central Asia and to Russia. Curiously, the beverage did not become popular in Central Asia proper until the sixteenth century, when it is mentioned in preserved texts from Bukhara and other areas.[70] This was likely due to the strict government restrictions on tea implemented by the Tea Horse Office in China, which mandated that all tea had to be sold to the government for resale to Tibet. In the sixteenth century, these regulations were loosened, opening new markets for tea. By 1638, when the German emissary Adam Olearius visited Isfahan, in modern-day Iran, he noted three types of taverns: brothels, coffee shops, and tea shops.[71]

After becoming established in Central Asia, tea spread far and wide. As with Buddhism in East and South Asia, the Muslim ban on alcohol boosted the popularity of tea as a stimulating drink. With the growth of the nautical trade in tea, consumption increased in Japan, the Islamic nations, and eventually in Europe and North America, where it exerted a strong influence on culture and altered political history.

CHAPTER THIRTEEN

Conclusion

With the price of East Asian spices soaring in Western Europe during the fifteenth century, the quest for black pepper, mace, and nutmeg led enterprising European sailors into unknown waters. Commissioned by Manuel the Fortunate of Portugal, Vasco da Gama, with his brother Paulo, set out in 1497 with a fleet of four ships around the Cape of Good Hope and across the Indian Ocean to Calicut. His journey, like the voyages of Christopher Columbus, changed the nature of global exchange forever. A study of the cultural impacts of the Age of Discovery raises the question of what the dinner tables of Europe looked like before the introduction of foods and spices from Asia and the New World.

Although Italian cuisine seems quintessentially European, many of its central ingredients were introduced to the Mediterranean only relatively recently. The tomato, domesticated in South America around three thousand years ago, was introduced to Europe as an oddity by Spanish explorers, and it took several centuries to become a popular food. Both pasta and the brick-oven flatbread that forms the basis of pizza were introduced to Italy from the Arab world, likely by medieval merchants. The flatbread was only a slight modification of the tandoori (or *tandir*) oven flatbreads baked in much of Asia, which may be topped with butter, herbs, and sauces. The Italians had only to add crushed tomatoes to create their national version. Likewise, the noodle, probably of East Asian origin, was introduced into the Mediterranean around a millennium ago by Arab merchants. Once the Italians of the late medieval period or early Renaissance got their hands on it, they appropriated it into their own cuisine and eventually topped it with crushed tomatoes as well.

Another Italian staple, polenta, is only a slight modification of the grain porridges that have been common across Europe since the Neolithic; but today it is almost invariably made with maize, a domesticate from southern Mexico. Gnocchi is an interesting Italian twist on the familiar dumpling. Today the dish is almost always prepared with *Solanum tuberosum*, a domesticated root from high in the Andes—the potato. Even the red pepper (*Capsicum* spp.) used to season Italian dishes and the chocolate (*Theobroma cacao*) in tiramisu are New World introductions, peppers having been domesticated roughly six thousand years ago in Mexico, and chocolate originating as an unsweetened beverage in Mesoamerica in the second millennium BC.

While it may be hard for some Italians to accept the idea that most of their cuisine was developed during the colonial era rather than in the banquet halls of ancient Rome, an even more surprising revelation is that modern Italian wine grapes grow not on centuries-old Italian vines but on rootstocks imported from North America. In the mid-nineteenth century the great European grape blight destroyed the vineyards of most European nations, hitting France the hardest. The blight was caused by grape phylloxera, an aphid (likely *Daktulosphaira vitifoliae*) that damaged the roots of the vines. The European wine industry came to a virtual standstill for two decades, until two French botanists discovered that grafting vines onto rootstocks from completely different North American grape species (initially roots from *Vitis aestivalis* grapes from Texas) gave the plants a level of immunity from the pest. During the 1870s and 1880s, the vineyards of Europe were gradually replanted with vines grafted onto North American blight-resistant rootstocks: thus all wine produced in Europe owes a debt to Texan grapes.

Other cuisines, too, are now largely dominated by introduced foods. In Central Asia, Russian imperialism introduced many new dishes, including *shchi* (cabbage soup) and borscht, as well as piroshki and blini. Even rice, which is the central ingredient of many Central Asian dishes today, can be traced back in most parts of the region only about 1,500 years (or possibly even less; see chapter 5).

This book is fundamentally concerned with the way globalization shaped cultures along the ancient Silk Road and continues to reshape the world around us. The journeys of Niccolò, Maffeo, and Marco Polo, as well as thousands of unnamed Magi, Sogdians, Persians, Uighurs, Gujaratis, Turks, and Arabs, all influenced the range of foods we eat today. As these travelers passed through the cities of Alexandria, Baghdad, Beirut, Bukhara, Constantinople, Damascus, Kafir Kala, Mecca, Muscat, Panjikent, Quanzhou, Samarkand, Thessaloniki, Turfan, Ulan Bator, and Xi'an, they picked up new plants and varieties of crops, which eventually dispersed to the far corners of the globe. One writer has observed that it is "only a slight exaggeration to call us servants of our food plants, diligently moving them around the world and slavishly tending them in manicured orchards and fields. And it's no exaggeration at all to call this activity seed dispersal."[1]

Caravan merchants and spice vendors were worldly, often multilingual, with well-honed social networking skills; they were adept at developing new markets and globalizing trade, and skilled at forging new alliances. With their caravans, they crossed not only deserts but also political barriers. As they did so, they carried the progenitors of plants that would eventually bear the legendary apricots of Damascus, the famous Hami melons, and the golden peaches of Samarkand. Prehistoric Central Asians also grew the small compote apples of Almaty, the large, juicy melons of Ashgabat and Samarkand, and a variety of grapes from Turfan with bright yellow skins, which yielded sweet red wines and dark purple raisins.

Figure 31. Photo taken between 1865 and 1872 of a Bactrian camel caravan in the Syr Darya region, carrying merchandise to markets. Photographer unknown. Library of Congress Prints and Photographs Division, Washington, DC.

The culinary magic that emanates from East Asian kitchens today is also the result of exotic ingredients, especially spices, introduced over millennia of trade. Frances Wood, the popular author of several books on the Silk Road and head of the Chinese section of the British Library, has observed that "aside from zoo animals and luxuries, foodstuffs were among the most important imports along the Silk Roads, for they greatly enlarged the potential of Chinese cuisine." She continues, "It would surprise many Chinese cooks to know that some of their basic ingredients were originally foreign imports. Sesame, peas, onions, coriander from Bactria, and cucumber were all introduced into China from the West during the Han dynasty."[2]

The Silk Road brought new ingredients to the kitchens of people all over the world, but it also had more profound effects on human history and agriculture. One of the most significant innovations brought about by the early movement of crops through Inner Asia was the rotation of crops. Based on archaeobotanical remains from Islamic villages in the Upper Euphrates, ranging in date from the eighth to the thirteenth centuries, one archaeobotanist has argued that farmers practiced a complex rotation of crops. The winter crops included two- and six-rowed hulled barley, free-threshing and glume wheat, rye, lentils, peas, chickpeas, and fava beans. The summer crops included cotton, rice, sesame, and both broomcorn and foxtail millet. He also identified a range of orchard and vineyard crops, as well as a few garden vegetables and herbs.[3]

Historical texts illustrate that a complex crop-rotation system was in place across Central Asia before Russian expansion. In 1821 and 1822, the explorer James Fraser noted the practice of crop rotation while crossing the Silk Road routes through Fergana, noting that it was similar to farming practices in the Zerafshan region. He observed that winter and summer crops were interchanged and mixed with orchards and cotton fields. He also noted that fruit was cultivated in the foothills and almonds, walnuts, and pistachios at higher elevations.[4] Eugene Schuyler, on his travels through the Fergana and Zerafshan regions in 1873, described a three-year rotation of winter wheat, barley, and one year of fallowing, and summer crops of rice, sorghum, cotton, flax, and vegetables.[5]

For most of human history winter was a much-needed period of repose for farmers, when crops did not grow. This allowed time for activities such as craft production and the development of social bonds. However, economic and population pressure gradually led to the planting of winter as well as summer crops, increasing productivity. In addition, the construction of irrigation works increased the productivity of the land but also demanded considerable additional human labor.

A similar process occurred with the introduction of drought-tolerant, fast-growing summer crops into arid regions. The spread of wheat into East Asia and broomcorn millet into West Asia and Europe, combined with the construction of centralized irrigation projects, changed human history forever. As Naomi Miller and her colleagues have pointed out, broomcorn millet gained importance in West Asia with the development of elaborate irrigation systems starting around 2,500 years ago.[6] With irrigation, a crop of millet could be grown in fields from which farmers had already harvested their winter wheat. Again, this crop rotation placed increasing demands on the soil and the farmers.

Likewise, wheat, introduced to China by the late third millennium BC, became a major winter crop in the Han period, with the construction of massive government-run dam and irrigation projects. In the same period, the *ard* plow was implemented and cast-iron blades were produced on a large scale for the first time. (Historians suggest that the plow was known in China long before this time and likely spread through Central Asia from southwest Asia.)[7] After this period, wheat was rotated with a summer rice crop. The increase in multicropping and the introduction of new milling technology from Central Asia likely led to the increased popularity of wheat during the Tang period, especially in the form of dumplings, fried dough, and noodles.[8] Central Asians in the Tang cities baked *shao-ping*, smaller versions of the naan flatbreads still produced in Central Asia today. Fermented dairy products also became increasingly popular at this time.

Intensive farming reached its peak in China with the fall of the Northern Song Dynasty (960–1126), when refugees migrating south brought their wheat-growing experience with them.[9] In addition, the Southern Song Dynasty (1131–63) collected taxes or rent on land based on the fall harvest only; therefore, farmers reaped tax-free profits from an additional spring or early summer harvest. The spread of fast-growing rice varieties into southern China allowed two rice crops in a year in the far south. In some areas of western China, barley became a winter crop in rotation with buckwheat.[10]

This increased agricultural productivity was a double-edged sword. The first generations of famers to switch to crop rotation may have profited handsomely, but the long-term effect was an increase in population, fed by the grain glut; a decrease in the value of the grain produced (necessitating larger harvests for farmers to make a living); and the rapid depletion of soils. Hence, the greater productivity and grain surplus enabled by crop rotation ultimately led to more arduous labor and oppression of agricultural workers and environmental degradation.

At the same time, the greater abundance of food freed part of the population from labor in the fields. This increase in leisure, which allowed time for specialized crafts and for education and study, ushered in the golden ages of art and innovation in both Asia and Europe. Typically, a food surplus is also invested into a militarized class, and this was the outcome of agricultural intensification across the Old World. This militarization inevitably amplified conflict: no one maintains a large standing army without using it. But even the military forces of the great powers reflected patterns of early botanical exchange through Central Asia: the Roman army was fed on unleavened millet bread and millet porridge, and the Mongolian mounted cavalry commanded by the Khans was fed on wheat-flour dumplings.[11]

For a century and a half the Silk Road was studied mainly for its effects on the empires and commercial centers of East Asia, South Asia, and the Mediterranean. With increased scientific investigation in Central Asia, however, scholars are now looking more closely at the prehistory and history of the Silk Road routes themselves. As new archaeological methods have been implemented and multidisciplinary excavations conducted, the old view of Central Asians as existing on the periphery of the ancient world has largely faded. The image of the steppe nomads as fierce warriors drinking their celebratory libations from the skulls of their fallen enemies (as Herodotus described them) has given way to a more nuanced understanding.[12] The Scythian, Saka, Wusun, and Xiongnu cultures were composed of a diverse range of people practicing mixed economic strategies, which included herding sheep and goats as well as growing several different crops. The history-book image of Scythian horsemen traversing thousands of kilometers across empty grasslands is gradually being replaced by the idea of small agropastoralist households connected to a broad social network.

In this book, we have crossed the routes of the Silk Road through the slopes and valleys of the mountainous spine of Central Asia. These grassy slopes fed camel caravans and seasonal transhumance herders for millennia; the glacial-melt streams watered fields and orchards, and the wild forests of Central Asia provided fruits, nuts, and game. These forests continued to play an important economic role into the twentieth century, especially from the Altai to the Pamir.[13] Although the low-growing shrubby forests in southern Central Asia, as well as in the river valleys of Zerafshan and Fergana, have been largely replaced by steppe vegetation, their importance is evident from the archaeobotanical data as well as from domesticated forms of the trees that once grew in these forests: pistachio, almond, *Elaeagnus*, *Crataegus*, and cherry, among others.[14] Humans have been manipulating these arboreal species and changing the composition of forests in this region for six thousand years. The apple pie and peach cobbler on your dinner table are artifacts not only of Silk Road trade but of the entire history of human settlement in Inner Asia.

While some historians, classicists, sinologists, and archaeologists still subscribe to the view that the Silk Road routes originated around the second century BC, archaeological data show that the process of exchange and interaction along these routes of the Silk Road began much earlier, by the late third millennium BC. The pattern of these early exchanges looks more like diffusion than organized interaction, but it is still part of the cultural phenomenon of Silk Road trade. The key to understanding the origins of organized exchange routes in Central Asia lies in the routes and methods devised to supply goods to high-altitude mining towns such as Sarazm starting in the fourth millennium BC. By the end of the first millennium BC, traders were moving goods through Central Asia along established routes, relying on the protection of military outposts funded by government taxation.

The first millennium BC saw great social and economic change across Inner Asia, resulting in what Nikolay Kradin refers to as a "complex pastoral society"[15] but also in greater investment in agricultural pursuits and the

Figure 32. View from the ruins of the ancient Silk Road city of Afrasiyab, which sat at the heart of the trade routes from at least the middle of the first millennium AD until 1220, when it was destroyed by the invading Mongols. In the background, the city of Samarkand is visible, with its Timurid madrasa, erected during the fifteenth and seventeenth centuries, standing above the city. Photo by the author.

introduction of new crop varieties. As the economist and agronomist Ester Boserup observed, population growth and increased social complexity are usually tied to the development or introduction of new agricultural technology and increased exchange. Therefore, we can envision the Silk Road not simply as a set of geographic routes but as a process of increased social connection across Eurasia, which ultimately fueled change and growing social complexity across the Old World. It became one of the most important conduits for the globalization of food, a corridor of exchange that has shaped cultures across Eurasia over the last five millennia.[16]

I close with the words of Owen Lattimore, the legendary explorer and scholar who traveled with the camel caravans of the Silk Road in the early

Figure 33. A vendor tending her wares in a market just outside Bukhara, Uzbekistan, 2018. She is selling dozens of varieties of fruits originating from across Uzbekistan and Tajikistan. Some vendors travel hundreds of kilometers to get fresh fruit to the markets. Photo by the author.

1920s, before the internal combustion engine and World War II transformed the trade routes. Writing in 1929, he lamented the changes that he had observed across western China:

> In our time, though every outward freight the caravans carry goes to work changes in Mongolia and Chinese Turkestan, they carry them in the old timeless way, as if no white men had ever come to Asia. Yet their doom is on them. When, in a time differed but inevitable,—for time in China goes by the half-century,—those who have learned from the foreigners shall have carried the railway through to Ning-hsia and Lan Chou, the business of caravans will decline into an affair of peddlers among the sand dunes of Alashan and the pastures of the Great Grasslands. It was a strange thing to walk in those markets, feeling the pulse of the life led through inenarrable yesterdays by the farthest-going caravans, and knowing the shadow of tomorrow would

distort all their type and character. When the camel man has done up his little bundle, he shambles away out of the city as if he were expecting to stroll home within half an hour; but he plods on until he finds the camp where the caravan waits behind the hills with its camels at pasture, until its complement of loads be filled; when camp is broken he plods away again until he fetches up in Central Asia; for the men of his calling, by leaving their houses and pitching tents, depart with no more ado from the civilization of telegraphs and newspapers, bayonets and martial law, into a secret and distant land of which they only know the doors.[17]

APPENDIX

European Travelers along the Silk Road

An excerpt from Eugene Schuyler's early travel accounts at Silk Road market towns, from a journey through Khiva, Tashkent, Samarkand, Bukhara, and Kokand in 1877:

> The gardens constitute the beauty of all this land. The long rows of poplar and elm trees, the vineyards, the dark foliage of the pomegranate over the walls, transport one at once to the plains of Lombardy or of Southern France. In the early spring the outskirts of the city, and indeed the whole valley, are one mass of white and pink, with the bloom of almond and peach, of cherry and apple, of apricot and plum, which perfume the air for miles around. These gardens are the favourite dwelling-places in the summer, and well may they be. Nowhere are fruits more abundant, and of some varieties it can be said that nowhere are they better. The apricots and nectarines I think it would be impossible to surpass anywhere. These ripen in June, and from that time until winter fruit and melons are never lacking. Peaches, though smaller in size, are better in flavour than the best of England, but they are far surpassed by those of Delaware. The big blue plums of Bukhara are celebrated through the whole of Asia. The cherries are mostly small and sour. The best apples come either from Khiva, or from Suzak, to the north of Turkistan, but the

small white pears of Tashkent are excellent in their way. The quince, as with us, is cultivated only for jams or marmalades, or for flavouring soup. Besides water-melons (tarbuz, whence the Russian arbuz) there are in common cultivation ten varieties of early melons, and six varieties which ripen later, any one of which would be a good addition to our gardens. In that hot climate they are considered particularly wholesome, and form one of the principal articles of food during summer. When a man is warm or thirsty, he thinks nothing of sitting down and finishing a couple of them. An acre of land, if properly prepared, would produce in ordinary years from two to three thousand, and in very good years twice as many. Of grapes I noticed thirteen varieties, the most of them remarkably good. The Jews distil a kind of brandy from the grapes, and the Russians have begun to make wine, but all the brands which I have seen, both red and white, were harsh and strong, and far inferior even to the wines of the Crimea or of the Caucasus. Large quantities of fruit are dried, and are known in Russian commerce by the name of izium or kishmish, although the latter is only properly applied to a certain variety of grape. If the fruit were dried properly and carefully, it might become a very important article of trade, as it is naturally so sweet that it can be made into compotes and preserves without the addition of sugar.[1]

Januarius MacGahan's 1876 description of the Bazaar of Khiva, from his accounts in *Campaigning on the Oxus and the Fall of Khiva:*

Out of this blinding glare you gladly step into the cool dark shade of the bazaar. A pleasant compound scent of spices, and many other agreeable odours, greet your nostrils; the confused noise and hum of a large crowd assail your ears; and an undistinguishable mass of men, horses, camels, donkeys, and carts meet your eyes. The bazaar is simply a street covered in, and it is altogether a very primitive affair. The roof is formed by beams laid from wall to wall across the narrow street, supporting small pieces of wood laid closely together, and covered with earth. It serves its purpose very well, however, and keeps out the heat and light. With delight you breathe the cool, damp, spice-laden air, and survey, with watering mouth, the heaps of rich, ripe fruit spread out in profusion. There are apricots, peaches, plums, grapes, and melons of a dozen different species, together with an indescribable array of wares only to be seen in Central Asia. Properly speaking, there are no shops; an elevated platform runs along one side, upon which men are seated among heaps of wares, with no apparent boundary line between them. On the other side there are a few barbers, butchers, cobblers, and smaller traders.

You push your horse with difficulty through the crowd for about fifty yards, until you come to another street, likewise covered in, which cuts this one transversely. Turning to the left, you enter a heavy arched gateway of brick, with massive wooden gates, and now you are in the "Tim," or bazaar proper. In this bazaar is transacted the principal retail business of the city. It is a double arcade or passage, 100 yards long, by forty feet wide; and is built of brick, in a succession of arches. The roof is some forty feet high, and each arch ends in a kind of dome-funnel, with a round hole in the top, which serves to light and ventilate the place. In the middle is a dome higher than the rest, which is not wholly without architectural pretensions.

The shops here are the merest booths or stalls, six or eight feet square, with one side open to the passage, displaying the most incongruous assortment of wares it is possible to imagine. Tea, sugar, silks, cotton-stuffs, khalats, boots, tobacco, everything, in short, found in Central Asia you will see displayed in one of these stalls. You take your seat in front of these booths, indulge your taste for fruit to its utmost on water-melons, cool and juicy, and peaches, rich and luscious, or grapes that make you think what a pity it is there is no wine. Or if you want to make a more substantial meal, a pilaoff with hot wheaten cakes, will be brought you in a twinkling; and you sit down there in the midst of the surging crowd, and quietly enjoy your meal. The tea used here, by the way, is the green, and is the only import, the supply of which is monopolised by the English.[2]

Part of Edmund O'Donovan's 1883 description of the Merv bazaar:

The bazaar consists of a labyrinth of narrow streets, lined on each side with the booths of the traders and artisans, in which the dealer arranges the commodities he has for sale, and behind which he sits, cross-legged, as a rule smoking the scarcely ever unlighted kalioun. The most numerous are the general dealers, who in addition to the orthodox tea, coffee, sugar, rice, and spices, also sell ink, paper, percussion caps, bullets, iron smallshot, gunpowder, brass drinking cups, salt, knives, sulphate of iron, pomegranate rind, alum for dyeing purposes, and an infinite variety of other articles. Turning a corner, we come into an alley where ropes suspended from housetop to housetop support numberless curtains of deep blue and olive-green calico. This is the quarter of the dyers, who seem to be, in point of number, the strongest after the bakhals, or grocers. They are to be seen working at their great indigo troughs, clad only in a dark-tinted waistband and skullcap, their arms, up to the elbows, being of as dark a blue as the calico which hangs outside. A little further on, towards the outskirts of the bazaar, are the vendors of fruit and vegetables, whose leeks and lettuces, spread in front of their booths, are a constant temptation to the passing camels and horses. More than once I have had to pay for the escapades of my horse in snatching up a bunch of spring onions and incontinently devouring it under the nose of the merchant. There were great basketsful of pomegranates and oranges, for Asterabad and its neighbourhood are famous for both these fruits, especially for the mandarin orange. Our ordinary orange is known as the portugal, while the naranj is quite as sour as any lemon, and takes the place of that fruit in cookery or with tea. Near the centre of the bazaar is a long street devoted to the coppersmiths, who manufacture tea-pots, saucepans, and cauldrons, for almost every cooking utensil used in this part of Persia is of copper, tinned inside, the facility of working copper more than compensating for the extra price of the material; moreover, the old vessels, when worn out, can be sold for a price very nearly equal to their cost when new. Now and then are to be seen cast iron pots of Russian manufacture, but these are much more in use among the Turcomans of the Atterek than in Persian households. The copper utensils are wrought by hand, and the din of hammering which salutes the ear as one enters the particular quarter of the smiths is perfectly deafening. By the sheer force of beating upon peculiar knob-like anvils, a hollow cylinder of copper, three-quarters of an inch in thickness, is made to expand to the most formidable dimensions. When finished, it is placed upon the fire, heated to dull redness, and a lump of tin is rubbed round inside.[3]

ACKNOWLEDGMENTS

I started working in Central Asia in graduate school, conducting archaeobotanical research under the direction of Michael Frachetti and Gayle Fritz at Washington University in St. Louis. I am extremely grateful for the intellectual foundation provided by Frachetti, Fritz, and Tristram R. Kidder; their diligent mentorship shines through in every page of this book. Inner Asia was largely an untouched region for the archaeological sciences, particularly archaeobotany. Over the past decade, however, interest in the spread of crops through this region has rapidly grown. Following my dissertation defense, I was awarded a Volkswagen and Mellon Foundations Postdoctoral Fellowship (Gr. CONF-673) to build on the research I had conducted in graduate school. While in Berlin on this fellowship, I was jointly affiliated with the the Eurasia Department of the German Archaeological Institute and the Paleontology Division of the Free University Berlin, working under the mentorship of Pavel Tarasov and Mayke Wagner. Many of the research ideas laid out in this book were developed during that year. Following my year of study in Berlin, I was awarded a Visiting Research Scholar fellowship through the Institute for the Study of the Ancient World (ISAW) and New York University, where I was afforded the time and freedom to finish writing this book in a highly collegial atmosphere.

I owe many of the ideas in this book to discussions with scholars at ISAW, including Roderick Campbell, Judith Lerner, Daniel Potts, Sören Stark, and

Karen Rubinson, among many others. Over the past several years I have also brainstormed with and received guidance from a number of scholars elsewhere. Just a few individuals who merit special note are Nicole Boivin, Claudia Chang, Paula Doumani Dupuy, Naomi Miller, Lynne Rouse, Farhad Maksudov, and Alicia Ventresca Miller. I extend special thanks to Yitzchak Jaffe, Mary Spengler, and Monica Yanda for reading chapters of the book and providing assistance with the development of ideas. I also acknowledge the hard work of Sören Stark, Marinda Brown, and a third anonymous reviewer, who read every page in detail and provided valuable feedback. This book reflects the intellectual contributions of many close colleagues.

NOTES

CHAPTER ONE. INTRODUCTION

1. Spengler III et al., 2014b.
2. Pollan, 2001.
3. Spengler III et al., 2014b; Spengler III, 2015.
4. Yunfei, Crawford, and Chen, 2015.
5. Huang et al., 2008.
6. Qian, 1993 [191–109 BC].
7. Jiang et al., 2009.
8. Rhindos, 1984.

CHAPTER TWO. PLANTS ON THE SILK ROAD

1. Spengler III, 2014.
2. Bābur, 1922 [1483–1530] volume 1, chapter 77.
3. Golden, 2011.
4. Bābur, 1922 [1483–1530], volume 1, chapter 92.
5. Jahangir, 1909–1914 [1569–1627], volume 1, chapter 73; volume 2, chapter 22.
6. Karev, 2013.
7. Golden, 2011.
8. Subtelny, 2013.

9. Karev, 2013.
10. Subtelny, 2013.
11. Mirsky, 1977.
12. Olearius, 2004 [1603–1671], 232.
13. Fazl, 1873–1907 [1597], volume 1, chapter 61.
14. Pellat, 1954.
15. Miquel, 1980; Samuel, 2001; Watson, 1983.
16. Sumner and Whitcomb, 1999.
17. El-Samarrahie, 1972.
18. Canard, 1959.
19. Ibn Fadlan, 2012, 88.
20. Ibn Fadlan, 2012, 177.
21. Ibn Hawqal, 1964.
22. Perry, 2017 [thirteenth century].
23. Chun, 1888 [1228].
24. Bembo, 2007 [1672].
25. Burnes, 1834, volume 2, 108, 168.
26. Fraser, 1825, xix, xx, 304.
27. Fraser, 1825, 432, 433, 752–53, 760.
28. Schuyler, 1877, 67.
29. MacGahan, 1876, 184, 198, 355, 313.
30. Landsell, 1885, 278, 498, 545, 605–9.
31. Schuyler, 1877.
32. Landsell, 1885, 498.
33. O'Donovan, 1883, 204, 252.
34. O'Donovan, 1883, 295–296.
35. O'Donovan, 1883, 106.
36. Maksudov et al., in press.
37. Boroffka et al., 2002; Spengler III and Willcox, 2013.
38. Spengler III et al., 2018.
39. Bubnova, 1987.
40. Bubnova, 1987.
41. Watson, 1983.
42. Laufer, 1919, 190.

43. Wood, 1998.
44. Vogel, 2013.
45. Polo, 1845 [ca. 1300].
46. Serventi and Sabban, 2002.
47. Anderson, 2014.
48. Watson, 1983.
49. Serventi and Sabban, 2002.

CHAPTER THREE. THE SILK AND SPICE ROUTES

1. Diffie and Winius, 1977.
2. Hanson, 2015.
3. Columbus, 2003 [1492], 128.
4. Hanson, 2015.
5. Birkill, 1953.
6. Harlan, 1971; Murdock, 1959.
7. Zeven and de Wet, 1982; Brunken, de Wet, and Harlan, 1977.
8. Fuller, 2003; Fuller and Boivin, 2009.
9. Nabhan, 2014.
10. Nabhan, 2014.
11. Singh, 2008.
12. Nabhan, 2014.
13. Watson, 1983.
14. Nabhan, 2014.
15. Walsh, 2006.
16. Liu, 2010.
17. Liu, 2010.
18. Nabhan, 2014.
19. Christian, 2000; Di Cosmo, 2002; Hanks and Linduff (eds.), 2009; Kuzmina, 2008; Victor (ed.), 2012.
20. Kuzmina, 2008, 64.
21. Gorbunova, 1986.
22. Christian, 2000, 2.
23. Richthofen, 1877.
24. Millward, 2013, 6.

25. Renfrew, 2014, xii.
26. Gorbunova, 1993.
27. Spengler III, 2015; Kuzmina, 2008; Renfrew, 2014.
28. Lattimore, 1928; Lattimore, 1940.
29. Doumani et al., 2015.
30. Christian, 2000; Rogers, 2007.
31. Qian, 1993 [109–91 BC], chapter 123.
32. Strabo, 1924 [7 BC–AD 23].
33. Hill, 2009.
34. Liu, 2010.
35. Mirsky, 1977.
36. Liu, 2010.
37. Anderson, 2014.
38. Anderson, 2014.
39. Anderson, 2014.
40. Wertmann, 2015; Anderson, 1988.
41. Wertmann, 2015; Anderson, 1988; Anderson, 2014.
42. Anderson, 2014.
43. Schafer, 1963.
44. Christian, 2000.
45. Hansen, 2003.
46. Karev, 2004.
47. Liu, 2010.
48. Liu, 2010.
49. Wertmann, 2015.
50. Liu, 2010.
51. Translated by Liu, 2010, 113.

CHAPTER FOUR. THE MILLETS

1. Herodotus, 1920 [ca. 450 BC], book 4, chapter 7; book 1, chapter 193, sections 1–5.
2. Miller, Spengler III, and Frachetti, 2016.
3. Valamoti and Jones, 2010.
4. Scarborough, 1978.

5. Dioscorides, 2000 [AD 64]; Osbaldeston, 2000.
6. Murphy, 2016.
7. Pliny the Elder, 1855 [AD 77–79], book 18, chapter 10; book 18, chapter 24; book 14, chapter 19; book 18, chapter 10.
8. Strabo, 1924 [7 BC–AD 23], book 7, chapter 5, section 5; book 5, chapter 1, section 3; book 5, chapter 4, section 4; book 12, chapter 3.
9. Hesiod, 1914 [eighth century BC], line 395.
10. Xenophon 1922 [ca. 365 BC], book 2, chapter 1, section 25.
11. Polybius, 1962 [ca. 140 BC], book 2.
12. Murphy, 2016.
13. Murphy, Thompson, and Fuller, 2013.
14. Murphy, 2016.
15. Zohary, Hopf, and Weiss, 2012.
16. Lisitsyna and Prishchepenko, 1977.
17. Lisitsina, 1984.
18. Hunt et al., 2008; Motuzaite-Matuzeviciute et al., 2013.
19. Zhao, 2011.
20. Harlan, 1975; Harlan, 1977.
21. Cho et al., 2010.
22. Hunt et al., 2011.
23. Hunt et al., 2008.
24. Boivin, Fuller, and Crowther, 2012.
25. Motuzaite-Matuzeviciute et al., 2013.
26. Bellwood, 2005; Crawford et al., 2005; Zhao, 2011.
27. Liu, Hunt, and Jones, 2009.
28. Bettinger et al., 2010; Liu, Kong, and Lang, 2004.
29. Zhao, 2011.
30. Barton et al., 2009.
31. Barton et al., 2009; Jing and Campbell, 2009.
32. Zhao, 2011.
33. Lu et al., 2009a.
34. Anderson, 2014; Liu, 2004; Yang et al., 2012.
35. Lu et al., 2009a.
36. Lu et al., 2009a, 7368–69; Lu et al., 2009b, 7, 12.

37. Lu et al., 2009a, 7367.
38. Yang et al., 2012.
39. Bestel et al., 2014; Zhao, 2011.
40. Liu et al., 2010; Yang et al., 2012.
41. Liu et al., 2010.
42. Fuller and Qin, 2009; Fuller and Qin, 2010; Fuller, Harvey, and Qin, 2007; Anderson, 2014; Liu et al., 2010.
43. Yang et al., 2012, 3726.
44. Wang et al., 2016, 6444.
45. Frachetti, 2012.
46. Spengler III et al., 2014b.
47. Miller, Spengler III, and Frachetti, 2016.
48. Trifonov et al., 2017.
49. Spengler III, 2015.
50. Paskhevich, 2003; Vainshtein, 1980.
51. Strabo, 1924 [7 BC–AD 23], book 5, chapter 1, section 12.
52. Pashkevich, 1984.
53. Motuzaite-Matuzeviviute, Telizhenko, and Jones, 2012.
54. Shantz and Piemeisel, 1927.
55. Miller, Spengler III, and Frachetti, 2016.
56. Di Cosmo, 1994; Levin and Potapov, 1964; Priklonskii, 1953 [1881]; Seebohm, 1882; Vainshtein, 1980.
57. Vainshtein, 1980.
58. Di Cosmo, 1994.
59. Lattimore, 1940.
60. Argynbaev, 1973, 155.
61. Bartol'd, 1962–1963.
62. Soucek, 2000.
63. Semple, 1928.
64. Columella, 1911 [mid-first century AD].
65. Hamd-Allāh Mustawfī of Qazwīn, 1919 [1340], chapter 22.
66. Samuel, 2001.
67. Ibn al-Awwam, 2000 [twelfth century].
68. Samuel, 2001; Ibn al-Awwam, 2000 [twelfth century].

69. Samuel, 2001.
70. Golden, 2011.
71. Fuller and Rowlands, 2011.
72. Murphy, 2016.
73. Murphy, 2016; Semple, 1928.
74. Miller, Spengler III, and Frachetti, 2016; Murphy, 2016.
75. Spengler III et al., 2014b.
76. Frachetti et al., 2010.
77. Miller, Spengler III, and Frachetti, 2016.
78. Spengler III et al., 2017b.
79. Spengler III et al., 2014a.
80. Rouse and Cerasetti, 2014.
81. Herrmann and Kurbansakhatov, 1994.
82. Costantini, 1979.
83. Nesbitt and Summers, 1988.
84. Spengler III, Chang, and Tortellotte, 2013.
85. Nesbitt, 1994; Wu, Miller, and Crabtree, 2015.
86. Helbaek, 1966; Neef, 1989.
87. Fuller and Boivin, 2009.
88. Tafuri, Graig, and Canci, 2009; Valamoti, 2016.
89. Riehl, 1999.
90. Fuller and Boivin, 2009; Spengler III et al., 2017b.
91. Trifonov et al., 2017.
92. Svyatko et al., 2013.
93. Spengler III et al., 2016.
94. Murphy et al., 2013.
95. Ventresca Miller et al., 2014.
96. Motuzaite-Matuzeviciute et al., 2015.
97. Miller, Spengler III, and Frachetti, 2016.
98. Zhao, 2011; Crawford et al., 2005.
99. Bestel et al., 2014.
100. de Wet, 1995.
101. Zohary, Hopf, and Weiss, 2012.
102. Miller, Spengler III, and Frachetti, 2016.

103. Hunt et al., 2008.
104. Spengler III, Doumani, and Frachetti, 2014.
105. Doumani et al., 2015.
106. Miller, 2010.
107. Miller, Spengler III, and Frachetti, 2016.
108. Wu, Miller, and Crabtree, 2015.
109. Spengler III, Chang, and Tortellotte, 2013.
110. Nesbitt and Summers, 1988.
111. Watson, 1983; Miller, Spengler III, and Frachetti, 2016.

CHAPTER FIVE. RICE AND OTHER ANCIENT GRAINS

1. Simoons, 1990.
2. Anderson, 1988, 140.
3. Anderson, 2014.
4. Anderson, 1988.
5. Anderson, 1988.
6. Simoons, 1990.
7. Ding, 1957; Chang, 1976; Higham, 1995.
8. See Gross and Zhao, 2014.
9. Stevens et al., 2016; Zhao, 2011.
10. Zheng et al., 2010; Crawford, 2006.
11. Sang and Ge, 2007a, 2007b; Vaughan, Lu, and Tomooka, 2008; Fuller, 2011, 2012.
12. Fuller, 2012.
13. Gross and Zhao, 2014.
14. Zhao, 2011.
15. Gross and Zhao, 2014.
16. Jiang and Liu, 2006; Anderson, 1988; Liu, 2004.
17. Fuller, Harvey, and Qin, 2007.
18. Fuller et al., 2009.
19. Zheng, Jiang, and Zheng, 2004.
20. Fuller et al., 2009.
21. Fuller et al., 2009; Gross and Zhao, 2014.
22. Civáň et al., 2015.

23. Tang et al., 2004.
24. Kawakami et al., 2007; Vaughan, Lu, and Tomooka, 2008.
25. Yang et al., 2012; Civáň et al., 2015.
26. McNally et al., 2009; Kovach, Sweeney, and McCouch, 2007; Sang and Ge, 2007.
27. Fuller and Qin, 2010; Fuller, 2011; Gross and Zhao, 2014; Stevens et al., 2016; Vaughan, Lu, and Tomooka, 2008.
28. Fuller, 2011; Stevens et al., 2016.
29. Costantini, 1987; Saraswat and Pokharia, 2003; Stevens et al., 2016.
30. Fuller, 2011; Silva et al., 2015; Fuller, 2012.
31. Silva et al., 2015.
32. Fuller, 2012; Stevens et al., 2016.
33. Fuller, 2012; Silva et al., 2015.
34. Stevens et al., 2016.
35. Bacon, 1980.
36. Watson, 1983.
37. Laufer, 1919.
38. Miller, 1981.
39. Nesbitt, Simpson, and Svanberg, n.d.
40. Miller, 1981.
41. Strabo, 1924 [7 BC–AD 23], book 15, chapter 1, section 18.
42. Nesbitt, Simpson, and Svanberg, n.d.
43. Siculus, 1967 [ca. 60 BC], book 2, chapter 36.
44. Theophrastus, 1916 [ca. 350–287 BC], book 4, section 4.
45. Dioscorides, 2000 [AD 64]; Pliny the Elder, 1855 [AD 77–79].
46. van der Veen, 2011.
47. Qian, 1993 [91–109 BC], book 123.
48. Laufer, 1919.
49. Gorbunova, 1986.
50. Bashtannik, 2008.
51. Rosen, Chang, and Grigoriev, 2000; Chang et al., 2002; Chang et al., 2003.
52. Spengler III, Chang, and Tortellotte, 2013.
53. Ghosh et al., 2008.
54. Song et al., 2018.

55. Lone, Khan, and Buth, 1993.
56. Knörzer, 2000; Lone, Khan, and Buth, 1993; Stevens et al., 2016.
57. Li et al., 2013.
58. Laufer, 1919.
59. Watson, 1983, 17.
60. Perry, 2017 [thirteenth century].
61. Watson, 1983.
62. Samuel, 2001.
63. El-Samarrahie, 1972.
64. Samuel, 2001.
65. Nesbitt, Simpson, and Svanberg, n.d.
66. Sumner and Whitcomb, 1999.
67. Anderson, 2014.
68. Nesbitt, Simpson, and Svanberg, n.d.
69. Samuel, 2001.
70. Bābur, 1922 [1483–1530], volume 2, chapter 14.
71. Fazl, 1873–1907 [1597], volume 1, chapters 49–55.
72. Fuller et al., 2009.
73. Gross and Zhao, 2014.
74. Stevens et al., 2016.
75. Simoons, 1990; Zeven and de Wet, 1982; Harlan, 1975.
76. Ohnishi, 2004.
77. Schuyler, 1877.
78. Chun, 1888 [1228].
79. Landsell, 1885.
80. Schuyler, 1877.
81. Janik, 2002.
82. Boivin, Fuller, and Crowther, 2012.
83. Hunt, Shang, and Jones, 2017.
84. de Candolle, 1884.
85. Crawford, 1983.
86. Ho, 1975.
87. Hunt, Shang, and Jones, 2017.
88. Xue, 2010.

89. Ohnishi, 1998.
90. Knörzer, 2000.
91. Hunt, Shang, and Jones, 2017.
92. Knörzer, 2000.
93. Song et al., in review.
94. d'Alpoim Guedes et al., 2014.
95. Asouti and Fuller, 2009.
96. Kimata, Ashok, and Seetharam, 2000.
97. Jiang et al., 2009.
98. Simoons, 1990.
99. Samuel, 2001.
100. Simoons, 1990.
101. Ho, 1975.
102. Pliny the Elder, 1855 [AD 77–79], book 18, chapter 10.
103. Simoons, 1990.
104. Danilevsky, Kokonov, and Neketen, 1940.

CHAPTER SIX. BARLEY

1. Grant, 2000.
2. Jones et al., 2008.
3. Lister and Jones, 2013.
4. Lister and Jones, 2013.
5. Willcox, 2013.
6. Harlan and Zohary, 1966.
7. Morrell, Lundy, and Clegg, 2003; Morrell and Clegg, 2007.
8. Harris, 2010; Anderson, 2014.
9. Morrell and Clegg, 2007; Harris, 2010.
10. Azguvel and Komatsuda, 2007; Takahashi, 1972.
11. Zohary, 1999.
12. Blattner and Méndez, 2001; Leon, 2010; Li, Xu, and Zhang, 2004.
13. Badr et al., 2000.
14. Molina-Cano et al., 1999; Molina-Cano et al., 2005; Xu, 1982.
15. Morrell, Lundy, and Clegg, 2003.
16. Xu, 1982; Ma et al., 1987.

17. Tanno and Takeda, 2004.
18. Yang et al., 2008.
19. Dai et al., 2012.
20. Langlie et al., 2014.
21. Snir and Weiss, 2014; Willcox, 2013.
22. Riehl, Zeidi, and Conard, 2013.
23. Weiss and Zohary, 2011.
24. Willcox, 2013.
25. Langlie et al., 2014.
26. Jones et al., 2013.
27. Jones et al., 2013.
28. Komatsuda et al., 2007; Leon, 2010.
29. Taketa et al., 2008.
30. Helbaek, 1959.
31. Takahashi et al., 1963; Takahashi et al., 1968.
32. Jones et al., 2008; von Bothmer et al., 2003.
33. Pourkheirandish and Komatsuda, 2007; von Bothmer et al., 2003; Takahashi et al., 1968; Takahashi et al., 1963.
34. Jones et al., 2008; Jones et al., 2012; Lister et al., 2009; von Bothmer et al., 2003.
35. Jones et al., 2008.
36. Lister and Jones, 2013.
37. Harris, 2010; Willcox, 1991; Spengler III and Willcox, 2013.
38. Miller, 1990.
39. Lister and Jones, 2013.
40. Harrison, 1995.
41. Spengler III et al., 2014a.
42. Spengler III, Doumani, and Frachetti, 2014.
43. Miller, 1999.
44. Spengler III et al., 2014.
45. Spengler III, Chang, and Tortellotte, 2013.
46. Wu, Miller, and Crabtree, 2015.
47. Wu, Miller, and Crabtree, 2015; d'Alpoim Guedes et al., 2014.
48. Knörzer, 2000.

49. Spengler III, 2015.
50. Spengler III and Willcox, 2013.
51. Spengler III et al., 2014a.
52. Tengberg, 1999; Willcox, 1994.
53. Costantini, 1987; Costantini, 1984.
54. Spengler III, Doumani, and Frachetti, 2014.
55. Flad et al., 2010; Jia, Betts, and Wu, 2011; Fu, 2001.
56. Miller, 2003, 130.
57. Burnes, 1834, volume 2, 244.
58. Lu et al., 2016.
59. Spengler III, 2015; Spengler III et al., 2014b; Spengler III et al., 2016.
60. Flad, 2017; Lu, 2016.
61. Quinn, Bista, and Childs, 2015; Simonson et al., 2010.
62. Morgan et al., 2011.
63. Aldenderfer, 2006; Aldenderfer and Zhang, 2004.
64. Ranov and Bubnova, 1961.
65. Simonson et al., 2010; Barton, 2016.
66. Bellwood, 2005.
67. Van Driem, 1999; Van Driem, 2002; Bellwood, 2005.
68. Brantingham and Xing, 2006.
69. Rhode et al., 2007.
70. Barton, 2016.
71. Aldenderfer and Zhang, 2004.
72. Li, 2007.
73. d'Alpoim Guedes et al., 2014.
74. Aldenderfer and Zhang, 2004; Li, 2007.
75. Li, 2007; d'Alpoim Guedes et al., 2014.
76. Chen et al., 2015.
77. Fu, 2001.
78. Asouti and Fuller, 2009.
79. Fu, 2001.
80. Chen et al., 2015.
81. Knörzer, 2000.
82. Knörzer, 2000.

83. Lone, Khan, and Buth, 1993; Sharma, 2000.
84. Wagner et al., 2011.
85. d'Alpoim Guedes et al., 2014.
86. Liu et al., 2016.
87. Aldenderfer, 2013.
88. Beckwith, 1993.

CHAPTER SEVEN. THE WHEATS

1. Bray, 1984.
2. Tanno and Willcox, 2012; Fuller et al., 2014; Riehl, Zeidi, and Conard, 2013.
3. Zohary, Hopf, and Weiss, 2012.
4. Dvořák et al., 1993; Dvořák et al., 1998.
5. Harris, 2010.
6. Zohary, Hopf, and Weiss, 2012.
7. Watson, 1983.
8. Crawford, 2006; Flad et al., 2010; Spengler III, 2015.
9. Crawford and Lee, 2003.
10. Li, Lister, and Li, 2011.
11. Samuel, 2001.
12. Spengler III et al., 2018.
13. Stevens et al., 2016; Zohary, Hopf, and Weiss, 2012; Kirleis and Fischer, 2014.
14. Salunkhe et al., 2012; Stevens et al., 2016; Fuller, 2002.
15. Stevens et al., 2016.
16. Saraswat and Pokharia, 2003; Saraswat and Pokharia, 2004.
17. Fuller, 2006; Saraswat, 1986; Saraswat and Pokharia, 2003; Saraswat and Pokharia, 2004; Stevens et al., 2016.
18. Zohary, Hopf, and Weiss, 2012.
19. Flad et al., 2010; Frachetti et al., 2010; Li et al., 2007; Zhao, 2009.
20. Barton and An, 2014; Betts, Jia, and Dodson, 2013; Dodson et al., 2013; Liu et al., 2016; Spengler III, 2015; Spengler III, Doumani, and Frachetti, 2014; Spengler III et al., 2014b; Spengler, 2013.
21. Zhao, 2009; see also Spengler III, 2015.

22. Ho, 1975.
23. Anderson, 2014.
24. Anderson, 1988.
25. Simoons, 1990.
26. Doumani et al., 2015; Spengler III, 2014.
27. Frachetti et al., 2010.
28. Doumani et al., 2015; Frachetti et al., 2010.
29. Hudaikov et al., 2013.
30. Jiang et al., 2009; Koroluyk and Polosmak, 2010.
31. Li et al., 2007.
32. Flad et al., 2010; Stevens et al., 2016.
33. Crawford et al., 2005.
34. Flad et al., 2010.
35. Spengler III, 2015; Spengler III et al., 2014b.
36. Liu et al., 2016.
37. Jiang et al., 2009; Wang, 1983.
38. Jia, Betts, and Wu, 2011.
39. Li et al., 2007.
40. Flad et al., 2010.
41. Li, Lister, and Li, 2011.
42. Dodson et al., 2013.
43. Flad et al., 2010.
44. Dodson et al., 2013.
45. Li et al., 2013.
46. Li et al., 2013.
47. Fu, 2001.
48. Xue, 2010.
49. Asouti and Fuller, 2009.
50. Lone, Khan, and Buth, 1993; Sharma, 2000.
51. Costantini, 1987.
52. Jacomet, 2006.
53. Miller, 1999; Spengler III, 2015; Spengler III et al., 2014a; Spengler III, Doumani, and Frachetti, 2014; Spengler III et al., 2014b.
54. Spengler, 2013.

55. Moore et al., 1994; Miller, 1999; Miller, 2003.
56. Moore et al., 1994.
57. Motuzaite-Matuzeviciute et al., 2016.
58. Costantini, 1984.
59. Weber, 1991; Vishnu-Mittre, 1972; Shaw, 1943.
60. Lunina, 1984.
61. Liu et al., 2016; Spengler III, 2015.
62. Fuller, 2001; Liu et al., 2016.
63. Kim, 2013; Braadbaart, 2008.
64. Kim, 2013.
65. Willcox, 1991, 146, 147.
66. Jacomet, 2006.
67. Spengler III, Doumani, and Frachetti, 2014; Spengler III et al., 2014a.
68. Spengler III et al., 2014b.
69. Spengler III, 2015; Fuller, 2001.
70. Peterson, 1965, 89.
71. Percival, 1921; Singh, 1946; Peterson, 1965; Rao, 1977.
72. Burt, 1941; Lone, Khan, and Buth, 1993; Stapf, 1931; Shaw, 1943; Vishnu-Mittre, 1972.
73. Costantini, 1984.
74. Miller, 2011.
75. Miller, 1999.
76. Spengler III and Willcox, 2013.
77. Casal, 1961; Hiebert, 2003; Hiebert, 1994; Kuzmina, 2008.
78. Crawford and Lee, 2003.
79. Kim, 2013.
80. Josekutty, 2008, unpublished.
81. Koba and Tsunewaki, 1978.
82. Salina et al., 2000.
83. Gegas et al., 2010.
84. Asakura et al., 2011.
85. Chen et al., 2012.
86. Salvi, 2013.
87. Reynolds and Borlaug, 2006.

88. Kim, 2013.
89. Chen et al., 2012; Borojevic and Borojevic, 2005.
90. Josekutty, 2008, unpublished.
91. Josekutty, 2008, unpublished.
92. Kim, 2013.
93. Salvi, 2013; Salvi, 2014.
94. Spengler III, 2015; Spengler III et al., 2014b.

CHAPTER EIGHT. LEGUMES

1. Spengler III et al., 2018; Spengler III et al., 2017a.
2. Lightfoot, Liu, and Jones, 2013.
3. Samuel, 2001.
4. Fuller et al., 2014; Riehl, Zeidi, and Conard, 2013; Tanno and Willcox, 2012.
5. Tanno and Willcox, 2006a.
6. Abbo et al., 2011.
7. Mirsky, 1977.
8. Zohary, Hopf, and Weiss, 2012.
9. Zohary, Hopf, and Weiss 2012.
10. Tanno and Willcox, 2006b.
11. Kislev, 1985.
12. Danilevsky, Kokonov, and Neketen, 1940.
13. Yakubov, 1979.
14. Danilevsky, Kokonov, and Neketen, 1940.
15. Zohary, Hopf, and Weiss, 2012.
16. Miller, 1999; Moore et al., 1994.
17. Moore et al., 1994.
18. Willcox, 1991.
19. Spengler III et al., 2014b.
20. Spengler III, Doumani, and Frachetti, 2014.
21. Weber, 1991.
22. Brite and Marston, 2013.
23. Spengler III et al., 2017b.
24. Spengler III et al., 2017b.

25. Laufer, 1919.
26. Laufer, 1919.
27. Simoons, 1990, 75.
28. Laufer, 1919.
29. Simoons, 1990; Laufer, 1919.
30. Fu, 2001; Spengler III, Doumani, and Frachetti, 2014.
31. Bashtannik, 2008.
32. Simoons, 1990, 149.
33. Brite and Marston, 2013.
34. Zeven and de Wet, 1982; Laufer, 1919.
35. Simoons, 1990.
36. de Candolle, 1884.
37. Strabo, 1924 [7 BC–AD 23], book xii, page 560; Pliny the Elder, 1855 [AD 77–79], book xvii, chapter 16.
38. de Candolle, 1884.
39. Fu, 2001.
40. Spengler III et al., 2014b.

CHAPTER NINE. GRAPES AND APPLES

1. Behr, 1981.
2. Strabo, 1924 [7 BC–AD 23], book 11, chapter 5, section 2.
3. Herodotus, 1920 [ca. 450 BC], book 1, chapter 133.
4. Strabo, 1924 [7 BC–AD 23], book 11, chapter 8, section 5.
5. Wertmann, 2015.
6. Stein, 1998 [1932].
7. Jiang et al., 2009.
8. Jiang et al., 2009.
9. Qian, 1993 [91–109 BC].
10. Huang, 2000; Huang et al., 2008.
11. Huang, 2000.
12. Yakubov, 1979.
13. Danilevsky, Kokonov, and Neketen, 1940.
14. Laufer, 1919.
15. Yakubov, 1979.

16. Laufer, 1919.
17. Hill, 2009, sections 2 and 17.
18. Schafer, 1963.
19. Sterckx, 2011.
20. Watt et al., 2004.
21. Mirsky, 1977.
22. Laufer, 1919.
23. Huang, 2000.
24. Wertmann, 2015.
25. Liu, 2010, 100.
26. Wertmann, 2015.
27. Wu, Ravens, and Missouri Botanical Gardens, 2006.
28. Jiang et al., 2009.
29. Myles et al., 2010.
30. Miller, 2008.
31. McGovern et al., 1996; Miller, 2008.
32. McGovern et al., 1996.
33. McGovern, 2003.
34. Harrison, 1995.
35. Moore et al., 1994.
36. Miller, 2008.
37. Lone, Khan, and Buth, 1993.
38. Wu, Miller, and Crabtree, 2015.
39. Gorbunova, 1986.
40. Spengler III et al., 2018.
41. Spengler III, Chang, and Tortellotte, 2013.
42. Fall, Falconer, and Lines, 2002; Sherratt, 1981; Sherratt 1983.
43. Gorbunova, 1986.
44. Gorbunova, 1986.
45. Yakubov, 1979.
46. Lippolis and Manassero, 2015.
47. Yakubov, 1979.
48. Lippolis and Manassero, 2015.
49. Lippolis and Manassero, 2015.

50. Livshits, 2015, 274.
51. Livshits, 2015, 272.
52. Pilipko, 2001.
53. Pilipko, 2001.
54. Lyre, 2012.
55. Lyre, 2012.
56. Samuel, 2001.
57. Ibn Hawqal, 1964; Miquel, 1980.
58. Pollan, 2001.
59. Hanson, 2015, 184–185; see also Pollan, 2001; Juniper and Mabberley, 2006.
60. Thoreau, 1862.
61. Harris, Robinson, and Juniper, 2002; Cornille et al., 2012; Cornille et al., 2014; Pollan, 2001.
62. Cornille et al., 2014, 59.
63. Cornille et al., 2014.
64. Li et al., 2013.
65. de Candolle, 1884.
66. Zohary, Hopf, and Weiss, 2012.
67. Ellison et al., 1978.
68. Renfrew, 1987.
69. Zohary, Hopf, and Weiss, 2012.
70. Miller, 1999.
71. Spengler III et al., 2018.
72. Spengler III et al., 2018.

CHAPTER TEN. OTHER FRUITS AND NUTS

1. Spengler III et al., 2018.
2. Andrianov, 2016 [1969].
3. Samuel, 2001.
4. Samuel, 2001.
5. Laufer, 1919.
6. Theophrastus, 1916 [ca. 350–287 BC].

7. Pliny the Elder, 1855 [AD 77–79], book 15, section 11, line 13, and book 12, section 7.
8. Su et al., 2015.
9. Faust and Timon, 1995.
10. Wang and Zhuang, 2001.
11. Li, 1970; Harlan, 1971; Simoons, 1990; Zeven and de Wet, 1982.
12. Wang and Zhuang, 2001.
13. Zheng, Crawford, and Chen, 2014.
14. Fuller, Harvey, and Qin, 2007.
15. Fuller et al., 2009.
16. Fuller, Harvey, and Qin, 2007.
17. Fuller and Zhang, 2007.
18. Zheng, Crawford, and Chen, 2014.
19. Huang et al., 2008.
20. Jiang et al., 2009.
21. Spengler III et al., 2018.
22. Lone, Khan, and Buth, 1993; Fuller and Madella, 2001; Stevens et al., 2016.
23. Schafer, 1963.
24. Simoons, 1990.
25. Simoons, 1990.
26. Bābur, 1922 [1483–1530], volume 2, chapter 51, and chapter 50.
27. Fazl, 1873–1907 [1597], volume 1, chapter 61.
28. Samuel, 2001.
29. Samuel, 2001.
30. Chehrābād Salt Mine Project, 2014.
31. Chehrābād Salt Mine Project, 2014.
32. Watkins, 1976.
33. Laufer, 1919.
34. Pliny the Elder, 1855 [AD 77–79]; Dioscorides, 2000 [AD 64].
35. Faust, Surányi, and Nyujtö, 1998.
36. de Candolle, 1884.
37. Fuller and Zhang, 2007.
38. Jiang et al., 2009.

39. Yakubov, 1979.
40. Danilevsky, Kokonov, and Neketen, 1940.
41. Gorbunova, 1986; Spengler III et al., 2018.
42. Watkins, 1976.
43. Pliny the Elder, 1855 [AD 77–79].
44. Spengler III et al., 2018.
45. Spengler III et al., 2017a.
46. Simoons, 1990.
47. Li, 1969.
48. de Candolle, 1884.
49. Kerje and Grum, 2000.
50. Sebastian et al., 2010.
51. Fuller, 2006.
52. Spengler III et al., 2018.
53. Brite and Marston, 2013.
54. Andrianov, 2016 [1969].
55. Chehrābād Salt Mine Project, 2014.
56. Nabhan, 2014.
57. Anderson, 2014.
58. Laufer, 1919.
59. Simoons, 1990.
60. de Candolle, 1884.
61. Simoons, 1990.
62. Gupta, 2004.
63. de Candolle, 1884.
64. Laufer, 1919; Schafer, 1963.
65. Li et al., 2013.
66. Schafer, 1963.
67. Zeven and de Wet, 1982.
68. Spengler III, Chang, and Tortellotte, 2013.
69. Spengler III et al., 2018.
70. Spengler III et al., 2018.
71. Dzhangaliev, Salova, and Turekhanova, 2003; Levin and Potapov, 1964.
72. Dzhangaliev, Salova, and Turekhanova, 2003.

73. Dzhangaliev, Salova, and Turekhanova, 2003; Levin and Potapov, 1964; Seebohm, 1882.
74. Dzhangaliev, Salova, and Turekhanova, 2003.
75. Seebohm, 1882.
76. Dzhangaliev, Salova, and Turekhanova, 2003.
77. Levin and Potapov, 1964; Seebohm, 1882.
78. Seebohm, 1882.
79. Dzhangaliev, Salova, and Turekhanova, 2003.
80. Levin and Potapov, 1964; Seebohm, 1882.
81. Spengler III and Willcox, 2013.
82. Harris, 2010.
83. Miller, 1999.
84. Ho, 1975.
85. Simoons, 1990.
86. d'Alpoim Guedes et al., 2014.
87. Simoons, 1990.
88. Strabo, 1924 [7 BC–AD 23], book 15, chapter 2, section 10; Theophrastus, 1916 [ca. 350–287 BC], book 9, chapter 4, section 7.
89. Laufer, 1919.
90. Miller, 1999.
91. Miller, 1999.
92. Spengler III and Willcox, 2013.
93. Spengler III et al., 2018.
94. Pollegioni et al., 2017.
95. Mani, 2008.
96. Pollegioni et al., 2017.
97. Miquel, 1980.
98. Landsell, 1885.
99. Fraser, 1825.
100. Stein, 1907; Mirsky, 1977.
101. Li et al., 2013.
102. Chehrābād Salt Mine Project, 2014.
103. Pollegioni et al., 2017, 9.
104. Laufer, 1919.

105. Laufer, 1919.
106. Fuller, Harvey, and Qin, 2007.
107. Tuncle, Nout, and Brimer, 1993; Nout, Tuncle, and Brimer, 1995.
108. Laufer, 1919; Schafer, 1963; Simoons, 1990.
109. Lone, Khan, and Buth, 1993; Fuller and Madella, 2001; Stevens et al., 2016.

CHAPTER ELEVEN. LEAFY VEGETABLES, ROOTS, AND STEMS

1. Simoons, 1990.
2. Arias et al., 2014.
3. Arias et al., 2014.
4. Chalhoub et al., 2014.
5. Arias et al., 2014.
6. Li, 1959.
7. Simoons, 1990.
8. Spengler, 2013.
9. Spengler III et al., 2018.
10. Spengler, Frachetti, and Fritz, 2013.
11. El Hadidi, 1984.
12. Dioscorides, 2000 [AD 64], book 2, 144, and book 3, 163–164.
13. Pliny the Elder, 1855 [AD 77–79].
14. Boulos, 1985.
15. Fowler and Mooney, 1990; Li, 1969.
16. Anderson, 2014.
17. Anderson, 1988.
18. Wang, 1907.
19. Anderson, 2014; Sterckx, 2011; Anderson, 1988.
20. Simoons, 1990.
21. de Candolle, 1884.
22. Laufer, 1919.
23. Anderson, 2014.
24. de Candolle, 1884.
25. Laufer, 1919.
26. Simoons, 1990.

27. Zeven and de Wet, 1982; Li, 1959.
28. Simoons, 1990.
29. Bruno, 2006.
30. Langlie et al., 2014.
31. Fritz and Smith, 1988.
32. Simoons, 1990.
33. Zeven and de Wet, 1982.
34. Sterckx, 2011.
35. Anderson, 2014.
36. Anderson, 1988.
37. Sterckx, 2011.
38. Anderson, 2014.
39. Anderson, 2014.
40. Lu et al., 2016.
41. Lu et al., 2016.
42. Yang et al., 2009.
43. Zhijun Zhao, unpublished lecture, 2008; Xue, 2010.
44. Spengler, 2013.
45. Spengler, Frachetti, and Fritz, 2013; Spengler III, 2014.
46. Anderson, 1952.
47. Anthony et al., 2005; Popova, 2006; Motuzaite-Matuzeviviute, Telizhenko, and Jones, 2012.
48. Bokovenkov, 2006.
49. Spengler III, 2015; Spengler III, Chang, and Tortellotte, 2013; Spengler III, Doumani, and Frachetti, 2014; Spengler, 2013; Spengler III et al., 2018.
50. Flad et al., 2010.
51. Shishlina, 2008.
52. Spengler III, unpublished report.
53. Fuller and Zhang, 2007.
54. Spengler III et al., 2014a; Spengler III et al., 2017a.
55. Helbaek, 1952.
56. Anthony et al., 2005.
57. Popova, 2006.
58. Anthony et al., 2005.

59. Anthony et al., 2005; Popova, 2006.
60. Popova, 2006.
61. Popova, 2006.
62. Helbaek, 1952, 221.
63. Helbaek, 1950.
64. Popova, 2006.
65. Boulos, 1985.
66. Gordyagin, 1892; Popov, 1803; Stefanovsky, 1893.
67. Gordyagin, 1892; Popov, 1803; Stefanovsky, 1893.
68. Brockhaus and Efron, 1890–1907.
69. Banga, 1957.
70. Banga, 1957.
71. Simoons, 1990; Banga, 1957.
72. Simoons, 1990.
73. Laufer, 1919.
74. Anderson, 2014.
75. Pliny the Elder, 1855 [AD 77–79].
76. Simoons, 1990.
77. Simoons, 1990.
78. Vainshtein, 1980; Priklonskii, 1953 [1881]; Vainshtein, 1980; Seebohm, 1882.
79. Simoons, 1990.
80. Anderson, 1988.
81. Li, 1959; Zeven and de Wet, 1982; Simoons, 1990.
82. Vainshtein, 1980; Seebohm, 1882; Priklonskii, 1953 [1881].
83. Vainshtein, 1980.
84. Priklonskii, 1953 [1881], section 31, 23.
85. Humphrey, Mongush, and Telengid, 1994; Mowart, 1970; Popov, 1966; Levin and Potapov, 1964.
86. Humphrey, Mongush, and Telengid, 1994; Levin and Potapov, 1964.
87. Levin and Potapov, 1964; Mowart, 1970; Popov, 1966; Vainshtein, 1980.
88. Vainshtein, 1980; Priklonskii, 1953 [1881].
89. Priklonskii, 1953 [1881]; Seebohm, 1882; Di Cosmo, 1994.

CHAPTER TWELVE. SPICES, OILS, AND TEA

1. Laufer, 1919.
2. Perry, 2017 [thirteenth century].
3. Laufer, 1919.
4. van der Veen, 2011.
5. Norwich, 1989.
6. de Candolle, 1884.
7. Nabhan, 2014.
8. Zohary, Hopf, and Weiss, 2012.
9. Kislev, 1988.
10. Zohary, Hopf, and Weiss, 2012.
11. Laufer, 1919.
12. Nabhan, 2014.
13. Laufer, 1919.
14. Nabhan, 2014.
15. Zohary, Hopf, and Weiss, 2012.
16. Fazl, 1873–1907 [1597], volume 2, chapter 38.
17. Simoons, 1990.
18. Zeven and de Wet, 1982.
19. Laufer, 1919.
20. Marinova and Riehl, 2009.
21. Nabhan, 2014.
22. Nabhan, 2014; Anderson, 2014.
23. Herodotus, 1920 [ca. 450 BC], book 3, chapters 107–182.
24. Nabhan, 2014.
25. van der Veen, 2011.
26. Nabhan, 2014; Wood, 2002.
27. Nabhan, 2014.
28. Samuel, 2001.
29. El-Samarrahie, 1972; Samuel, 2001.
30. Zohary, Hopf, and Weiss, 2012.
31. Doumani, Spengler, and Frachetti, 2017.
32. Doumani, Spengler, and Frachetti, 2017.
33. Willcox, 1991.

34. Tengberg, 1999.
35. Costantini, 1979.
36. Fuller, 2011; Weber, 1991; Fuller, 2008.
37. Spengler III et al., 2014a; Spengler III et al., 2017b.
38. Knörzer, 2000.
39. Costantini, 1987.
40. d'Alpoim Guedes et al., 2015.
41. d'Alpoim Guedes et al., 2015.
42. Laufer, 1919.
43. Dinç et al., 2009; Jones and Valamoti, 2005.
44. Dinç et al., 2009.
45. Jones and Valamoti, 2005.
46. Spengler III et al., 2017b.
47. Jones and Valamoti, 2005.
48. Valamoti and Jones, 2010.
49. Mair and Hoh, 2009.
50. Mair and Hoh, 2009.
51. Fuchs, 2008.
52. Mair and Hoh, 2009.
53. Mair and Hoh, 2009.
54. Mair and Hoh, 2009.
55. Anderson, 1988.
56. Mair and Hoh, 2009.
57. Mair and Hoh, 2009.
58. Lu et al., 2016.
59. Yang, 2005.
60. Yang, 2005.
61. Yang, 2005.
62. Mair and Hoh, 2009.
63. Mair and Hoh, 2009.
64. Fuchs, 2008.
65. Fuchs, 2008.
66. Fuchs, 2008.
67. Selens and Freeman, 2011.

68. Mair and Hoh, 2009.
69. Liu, 2010.
70. Mair and Hoh, 2009.
71. Olearius, 2004 [1603–1671].

CHAPTER THIRTEEN. CONCLUSION

1. Hanson, 2015, 184–185.
2. Wood, 2002, 59.
3. Samuel, 2001.
4. Fraser, 1825.
5. Schuyler, 1877.
6. Miller, Spengler III, and Frachetti, 2016.
7. Anderson, 2014.
8. Anderson, 1988.
9. Bray, 1984.
10. Anderson, 1988.
11. Herodotus, 1920 [ca. 450 BC].
12. Dzhangaliev, Salova, and Turekhanova, 2003.
13. Spengler III and Willcox, 2013.
14. Li, 2002; Linduff, 2006; Spengler III, 2015.
15. Koryakova and Epimakhov, 2007; Kradin, 2002.
16. Spengler III, 2015; Stevens et al., 2016; Jones et al., 2011; Boivin, Fuller, and Crowther, 2012; Boivin et al., 2014; Frachetti, 2012.
17. Lattimore, 1995 [1929], 27.

APPENDIX

1. Schuyler, 1877, 296–297.
2. MacGahan, 1876, 303–305.
3. O'Donovan, 1883, 55–56.

REFERENCES

Abbo, S., E. Rachamin, Y. Zehavi, I. Zezak, S. Lev-Yadun, and A. Gopher. 2011. "Experimental growing of wild pea in Israel and its bearing on Near Eastern plant domestication." *Annals of Botany* 107: 1399–1404.

Aldenderfer, M. S. 2006. "Modelling plateaux peoples: The early human use of the world's high plateaux." *World Archaeology* 38 (3): 357–370.

Aldenderfer, M. S. 2013. "Variation in mortuary practice on the early Tibetan plateau and the high Himalayas." *Journal of the International Association for Bon Research* 1: 293–318.

Aldenderfer, M. S., and Yinong Zhang. 2004. "The prehistory of the Tibetan plateau to the seventh century A.D.: Perspectives and research from China and the west since 1950." *Journal of World Prehistory* 18 (1): 1–55.

Anderson, Edgar. 1952. *Plants, Man and Life*. Boston: Little Brown and Company.

Anderson, Eugene. 1988. *The Food of China*. New Haven, CT: Yale University Press.

Anderson, Eugene N. 2014. *Food and Environment in Early and Medieval China*. Philadelphia, PA: University of Pennsylvania Press.

Andrianov, B. V. 2016 [1969]. *Ancient Irrigation Systems of the Aral Sea Area*. Oxford: Oxbow Books.

Anthony, D. W., D. Brown, E. Brown, A. Goodman, A. Kokhlov, P. Kosintsev, O. Mochalov, et al. 2005. "The Samara Valley project: Late Bronze Age economy and ritual in the Russian steppes." *Eurasia Antiqua* 11: 395–417.

Argynbaev, Kh. 1973. "On the agriculture of the Kazakhs of Kopal Yesd, Semireche Oblast." In *Essays on the Agricultural History of the Peoples of Central Asia and Kazakhstan*, 154–160. Leningrad: Nauka.

Arias, Tatiana, Mark A. Beilstein, Michelle Tang, Michael R. McKain, and Chris Pires. 2014. "Diversification times among Brassica (Brassicaceae) crops suggest hybrid formation after 20 million years of divergence." *American Journal of Botany* 101 (1): 86–91.

Asakura, N., N. Mori, C. Nakamura, and I. Ohtsuka. 2011. "Comparative nucleotide sequence analysis of the D genome-specific sequence-tagged-site locus A1 in *Triticum aesticum* and its implication for the origin of subspecies *sphaerococcum*." *Breeding Science* 61: 212–216.

Asouti, Eleni, and Dorian Q. Fuller. 2009. "Archaeobotanical evidence." In *Grounding Knowledge, Walking Land: Archaeological Research and Ethnohistoric Identity in Nepal*, edited by Christopher Evans, Judith Pettigrew, Yarjung Kromchai, and Mark Turin, 142–152. London: Macdonald Insitute Monographs.

Azguvel, P., and T. Komatsuda. 2007. "A phylogenetic analysis based on nucleotide sequence of a marker linked to the brittle rachis locus indicates a diphyletic origin of barley." *Annual Journal of Botany* 100: 1009–1015.

Bābur. 1922 [1483–1530]. *The Bābur-Nāma in English (Memoirs of Bābur)*. Translated by Annette Susannah Beveridge. London: Luzac.

Bacon, Elizabeth Emaline. 1980. *Central Asia under Russian Rule: A Study in Culture Change*. Ithaca, NY: Cornell University Press.

Badr, A., K. Muller, R. Schafer-Pregl, H. El Rabey, S. Effgen, H. H. Ibrahim, C. Pozzi, W. Rohde, and F. Salamini. 2000. "On the origin and domestication history of barley (*Hordeum vulgare*)." *Molecular Biology and Evolution* 17: 499–510.

Banga, O. 1957. "Origin of the European cultivated carrot." *Euphytica* 6: 54–63.

Bartol'd, Vasiliĭ Vladimirovich. 1962–1963. *Four Studies on the History of Central Asia*. Translated by V. T. Minorsky. Leiden: E. J. Brill.

Barton, L., and C.-B. An. 2014. "An evaluation of competing hypotheses for the early adoption of wheat in East Asia." *World Archaeology* 46: 775–798.

Barton, Loukas. 2016. "The cultural context of biological adaptation to high elevation Tibet." *Archaeological Research in Asia* 5: 4–11.

Barton, Loukas, Seth D. Newsome, Fa-Hu Chen, Hui Wang, Thomas P. Guilderson, and Robert L. Bettinge. 2009. "Agricultural origins and the isotopic identity of domestication in northern China." *Proceedings of the National Academy of Sciences* 106 (14): 5523–5528.

Bashtannik, S. V. 2008. "Archaeobotanical studies at medieval sites in the Arys river valley." *Archaeology Ethnology and Anthropology of Eurasia* 33 (1): 85–92.

Beckwith, Christopher I. 1993. *The Tibetan Empire in Central Asia: A History of the Struggle for Great Power among Tibetans, Turks, Arabs, and Chinese during the Early Middle Ages*. Princeton: Princeton University Press.

Behr, Charles A., trans. 1981. *P. Aelius Aristides: The Complete Works*. Leiden: Brill.

Bellwood, P. 2005. *First Farmers: The Origins of Agricultural Societies*. Malden, MA: Blackwell.

Bembo, Ambrosio. 2007 [1672]. *The Travels and Journal of Ambrosio Bembo*. Berkeley: University of California Press.

Bestel, Sheahan, Gary W. Crawford, Li Liu, Jinming Shi, Yanhua Song, and Xingcan Chen. 2014. "The evolution of millet domestication, Middle Yellow River Region, North China: Evidence from charred seeds at the late Upper Paleolithic Shizitan Locality 9 site." *The Holocene* 24 (3): 261–265.

Bettinger, Robert, Loukas Barton, Christopher Morgan, Fahu Chen, Hui Wang, Thomas P. Guilderson, Duxue Ji, and Dongju Zhang. 2010. "The transition to agriculture at Dadiwan, People's Republic of China." *Current Anthropology* 51: 703–714.

Betts, A., P. W. Jia, and J. Dodson. 2013. "The origins of wheat in China and potential pathways for its introduction: A review." *Quaternary International* 30: 1–11.

Birkill, I. H. 1953. "Habits of man and the origins of the cultivated plants in the old world." *Proceedings of the Linnaean Society of London* 164: 12–42.

Blattner, F. R., and A. G. B. Méndez. 2001. "(2001) RAPD data do not support a second centre of barley domestication in Morocco." *Genetic Resources in Crop Evolution* 48: 13–19.

Bobrow-Strain, Aaron. 2013. *White Bread: A Social History of the Store-Bought Loaf*. Boston: Beacon Press.

Boivin, N., A. Crowther, M. Prendergast, and D. Q. Fuller. 2014. "Indian Ocean food globalization and Africa." *African Archaeological Review* 31 (4): 547–581.

Boivin, N., D. Q. Fuller, and A. Crowther. 2012. "Old World globalization and the Columbian exchange: Comparison and contrast." *World Archaeology* 44 (3): 452–469.

Bokovenkov, N. 2006. "The emergence of the Tagar culture." *Antiquity* 80: 860–879.

Boroffka, N., J. Cierny, J. Lutz, H. Parzinger, E. Pernicka, and G. Weisgerber. 2002. "Bronze Age tin from Central Asia: Preliminary notes." In *Ancient Interactions: East and West in Eurasia*, edited by K. Boyle, C. Renfrew, and M. Levine, 135–159. Oxford: McDonald Institute Monographs.

Borojevic, K., and K. Borojevic. 2005. "The transfer and history of 'reduced height genes' (*Rht*) in wheat from Japan to Europe." *Journal of Heredity* 96 (4): 455–459.

Boulos, Loutfy. 1985. "The Middle East." In *Plant Resources of Arid and Semiarid Lands: A Global Perspective*, edited by J. R. Goodin and David K. Northington, 187–232. New York: Academic Press.

Braadbaart, F. 2008. "Carbonization and morphological changes in modern dehusked and husked *Triticum dicoccum* and *Triticum aestivum* grains." *Vegetation History and Archaeobotany* 17: 155–166.

Brantingham, Jeffrey, and Gao Xing. 2006. "Peopling of the Northern Tibetan Plateau." *World Archaeology* 38: 387–414.

Bray, Francesca. 1984. "Agriculture." In *Science and Civilization in China, Part 2, Volume 6: Biology and Biological Technology*, edited by Joseph Needham. Cambridge: Cambridge University Press.

Brite, Elizabeth Baker, and John M. Marston. 2013. "Environmental change, agricultural innovation, and the spread of cotton agriculture in the Old World." *Journal of Anthropological Archaeology* 32: 39–53.

Brockhaus, F. A., and E. A. Efron. 1890–1907. *Энциклопедический словарь Брокгауза и Ефрона* (Brockhaus and Efron encyclopedic dictionary). St. Petersburg: Ilya Efron.

Brunken, Jere, J. M. de Wet, and J. R. Harlan. 1977. "The morphology and domestication of pearl millet." *Economic Botany* 31: 163–174.

Bruno, M. C. 2006. "A morphological approach to documenting the domestication of *Chenopodium* in the Andes." In *Documenting Domestication: New Genetic and Archaeological Paradigms*, edited by M. A. Zeder, D. G. Bradley, E. Emshwiller, and B. D. Smith, 32–45. Berkeley: University of California Press.

Bubnova, M. A. 1987. "*К Вопросу о Земледелии На Западном Памире В* IX-XI *вв* (Regarding the question of agriculture in the Western Pamirs from the 9th to the 11th century)." In *Прошлое Средней Азии: Археология, Нумизматика и Эпиграфика, Этнографния* (Past Central Asia: Numismatics, Archaeology and Ethnography Epigraphy), 59–66. Dushanbe, Tajikistan: Donish.

Burnes, Alex. 1834. *Travels into Bokhara: Being the account of a journey from India to Cabool, Tartary, and Persia*. London: John Murray.

Burt, B. C. 1941. "Comments on cereals and fruits." In *Excavations at Harappa*, edited by M. S. Vats, 466. New Delhi: Government of India Publications.

Canard, M. 1959. "Le riz dans le Proche Orient aux premiers siècles de l'Islam." *Arabica* 6: 113–131.

Casal, Jean-Marie. 1961. *Fouilles de Mundigak* (Mundigak excavations). Paris: Librairie C. Klincksieck.

Chalhoub, Boulos, France Denoeud, Shengyi Liu, Isobel A. Parkin, Haibao Tang, Xiyin Wang, Julien Chiquet, et al. 2014. "Early allopolyploid evolution in the post-Neolithic *Brassica napus* oilseed genome." *Science* 345 (6199): 950–953.

Chang, C., N. Benecke, F. P. Grigoriev, A. M. Rosen, and P. A. Tourtellotte. 2003. "Iron Age society and chronology in southeast Kazakhstan." *Antiquity* 77: 298–312.

Chang, Claudia, P. Tourtellotte, K. M. Baipakov, and F. P. Grigoriev. 2002. *The Evolution of Steppe Communities from Bronze Age through Medieval Periods in Southeastern Kazakhstan (Zhetysu)*. Sweet Briar, VA: Sweet Briar College.

Chang, T. 1976. "Rice." In *Evolution of Crop Plants*, edited by N. W. Simmonds, 98–104. London: Longman.

Chehrābād Salt Mine Project. 2014. *Extended Research Report on the Chehrābād Salt Mine Project 2010–2013*. Bochum: Deutsches Bergbau-Museum Bochum.

Chen, F. H., G. H. Dong, D. J. Zhang, X. Y. Liu, X. Jia, C. B. An, M. M. Ma, et al. 2015. "Agriculture facilitated permanent human occupation of the Tibetan Plateau after 3600 B.P." *Science* 347 (6219): 248–250.

Chen, G., Q. Zheng, Y. Bao, S. Liu, H. Wang, and X. Li. 2012. "Molecular cytogenetic identification of a novel dwarf wheat line with introgressed *Thinopyrum ponticum* chromatin." *Journal of Biological Science* 37 (1): 149–155.

Cho, Young-Il, Jong-Wook Chung, Gi-An Lee, Kyung-Ho Ma, Anupam Dixit, Jae-Gyun Gwag, and Yong-Jin Park. 2010. "Development and

characterization of twenty-five new polymorphic microsatellite markers in proso millet (*Panicum miliaceum L.*)." *Genes and Genomics* 32: 267–273.
Christian, D. 2000. "Silk Roads or steppe roads? The Silk Roads in world history." *Journal of World History* 11 (1): 1–26.
Chun, Chang. 1888 [1228]. *The Travels To the West of Kiu Ch'ang Ch'un, 1220–1223*. In *Medieval Researches from Eastern Asiatic Sources*, vol. 1, translated by Emil Bretschneider. London: Trubner & Co.
Civáň, P., H. Craig, C. J. Cox, and T. A. Brown. 2015. "Three geographically separate domestications of Asian rice." *Nature Plants* 1: 15164.
Cleary, Michelle Negus. 2013. "Khorezmian walled sites of the seventh century BC–fourth century AD: Urban settlements? Elite strongholds? Mobile centres?" *Iran* 1: 71–100.
Columbus, Christopher. 2003 [1492]. *Journal of the First Voyage of Columbus*. American Journeys Collection, Document No AJ-062. Madison: Wisconsin Historical Society.
Columella, L. J. M. 1911 [mid-first century AD]. Cambridge, MA: Harvard University Press.
Cornille, A., P. Gladieux, M. J. M. Smulders, I. Rold Án-Ruiz, F. Laurens, B. Le Cam, and A. Nersesyan. 2012. "New insight into the history of domesticated apple: Secondary contribution of the European wild apple to the genome of cultivated varieties." *PLoS Genetics* 8: e1002703.
Cornille, A., T. Giraud, M. J. Smulders, I. Rold Án-Ruiz, and P. Gladieux. 2014. "The domestication and evolutionary ecology of apples." *Trends in Genetics* 30: 57–65.
Costantini, L. 1979. "Plant remains at Pirak." In *Fouilles de Pirak*, edited by J.-F. Jarrige and M. Saontoni, 326–333. Paris: Boccard.
Costantini, L. 1984. "The beginning of agriculture in the Kachi Plain: The evidence of Mehrgarh." In *South Asian Archaeology 1981*, edited by B. B. Allchin, 29–33. Cambridge: Cambridge University Press.
Costantini, L. 1987. "Appendix B: Vegetal remains." In *Prehistoric and Protohistoric Swat, Pakistan*, edited by G. Stacul, 155–165. Rome: Instituto Italiano per il Medio ed Estremo Orientale.
Crawford, G. W. 2006. "East Asian plant domestication." In *Archaeology of Asia*, edited by M. T. Stark, 77–95. Oxford: Blackwell.

Crawford, Gary W. 1983. *Paleoethnobotany of the Kameda Peninsula Jomon*. Anthropological Papers No. 73, Museum of Anthropology, University of Michigan.

Crawford, G. W., and G.-A. Lee. 2003. "Agricultural origins in the Korean Peninsula." *Antiquity* 77: 87–95.

Crawford, Gary, Anne Underhill, Zhijun Zhao, Gyoung-Ah Lee, Gary Feinman, Linda Nicholas, Fengshi Luan, Haiguang Yu, Hui Fang, and Fengshu Cai. 2005. "Late Neolithic plant remains from northern China: Preliminary results from Liangchengzhen, Shandong." *Current Anthropology* 46 (2): 309–328.

d'Alpoim Guedes, J., H. Lu, Y. Li, X. Wu, R. Spengler, and M. Aldenderfer. 2014. "Moving agriculture onto the Tibetan Plateau: The archaeobotanical evidence." *Journal of Archaeological and Anthropological Science* 6: 255–269.

d'Alpoim Guedes, Jade A., Hongliang Lu, Anke M. Hein, and Amanda H. Schmidt. 2015. "Early evidence for the use of wheat and barley as staple crops on the margins of the Tibetan Plateau." *Proceedings of the National Academy of Sciences* 10 (1073): 1423708112.

Dai, F., E. Nevo, D. Wu, M. Comadran, L. Zhou, L. Qiu, Z. Chen, A. Beiles, G. Chen, and G. Zhang. 2012. "Tibet is one of the centers of domestication of cultivated barley." *Proceedings of the National Academy of Sciences* 109 (42): 16969–16973.

Danilevsky, V. V., V. N. Kokonov, and V. A. Neketen. 1940. "A study of the plant remains excavated from the eighth-century settlement of Mug in Tajikistan." In *Vegetation of Tajikistan and human interaction*, 484. Dushanbe: Tajikistan Base of Science of the SSSR. (In Russian.)

de Candolle, Alphonse. 1884. *Origin of Cultivated Plants*. London: K. Paul, Trench.

de Wet, J. M. J. 1995. "Foxtail millet." In *Evolution of Crop Plants*. 2nd ed., edited by J. Smartt and N. W. Simmonds, 170–172. London: Longman.

Di Cosmo, Nicola. 1994. "Ancient Inner Asian nomads: Their economic basis and its significance in Chinese history." *Journal of Asian Studies* 53 (4): 1092–1126.

Di Cosmo, Nicola. 2002. *Ancient China and Its Enemies: The Rise of Nomadic Power in East Asian History*. Cambridge: Cambridge University Press.

Diffie, Bailey W., and George D. Winius. 1977. *Foundations of the Portuguese Empire, 1415–1850: Europe and the World in the Age of Expansion*. Minneapolis: University of Minnesota Press.

Dinç, M., N. Münevver Pinar, S. Dogu, and Ş. Yildirimli. 2009. "Micromorphological studies of *Lallemantia L.* (Lamiaceae) species growing in Turkey." *Acta Biologica Cracoviesia Series Botanica* (1) 51 (1): 45–54.

Ding, Y. 1957. "The origin and evolution of Chinese cultivated rice." *Journal of Agriculture* 8 (3): 243–260.

Dioscorides, Pedanius. 2000 [AD 64]. *De Materia Medica* (On Medical Materials), edited and with an introduction by Tess Anne Osbaldeston. Johannesburg: Ibidis Press.

Dodson, J. R., X. Li, X. Zhou, K. Zhao, N. Sun, and P. Atahan. 2013. "Origin and spread of wheat in China." *Quaternary Science Reviews* 72: 108–111.

Doumani, Paula N., Michael D. Frachetti, R. Beardmore, T. Schmaus, Robert N. Spengler, and A. N. Mar'yashev. 2015. "Burial ritual, agriculture, and craft production among Bronze Age pastoralists at Tasbas (Kazakhstan)." *Archaeological Research in Asia* 1–2: 17–32.

Doumani, Paula, Robert N. Spengler, and Michael Frachetti. 2017. "Eurasian textiles: Case studies in exchange during the incipient and later Silk Road periods." *Quaternary International*, online first: 1–12.

Dvořák, J., M. C. Luo, Z. I. Yang, and H. B. Zhang. 1998. "The structure of the *Aegilops tauschii* genepool and the evolution of hexaploid wheat." *Theoretical and Applied Genetics* 67: 657–670.

Dvořák, J., P. di Terlizzi, H. B. Zhang, and P. Resta. 1993. "The evolution of polyploidy wheats: Identification of a genome donor species." *Genome* 36: 21–31.

Dzhangaliev, A. D., T. N. Salova, and P. M. Turekhanova. 2003. "The wild fruit and nut plants of Kazakhstan." In *Horticultural Reviews*, edited by Jules Janick, 305–372. New York: John Wiley.

El Hadidi, Nabil M. 1984. "Food plants of prehistoric dynastic Egypt." In *Plants for Arid Lands*, edited by G. E. Wickens, J. P. Goodin, and D. V. Field, 87–92. London: George Allen and Unwin.

Ellison, R., J. Renfrew, D. Brothwell, and N. Seeley. 1978. "Some food offerings from Ur, excavated by Leonard Woodley, and previously unpublished." *Journal of Archaeological Science* 5: 167–177.

El-Samarrahie, Q. 1972. *Agriculture in Iraq during the 3rd Century*. Beirut: A. H. Librairie du Liban.

Fall, Patricia, Steven E. Falconer, and Lee Lines. 2002. "Agricultural intensification and the secondary products revolution along the Jordan Rift." *Human Ecology* 4 (30): 445–482.

Faust, M., and B. Timon. 1995. "Origin and dissemination of peach." *Horticultural Review* 17: 331–379.

Faust, M., D. Surányi, and F. Nyujtö. 1998. "Origin and dissemination of apricot." *Horticultural Review* 22: 225–266.

Fazl, Abul Allámi. 1873–1907 [1597]. *The Aín I Akbari*. Translated by H. Blochmann and H.S. Jarrett. Calcutta: Asiatic Society of Bengal and the Baptist Mission Press.

Flad, Rowan. 2017. "Recent research on the archeology of the Tibetan Plateau and surrounding areas." *Archaeological Research in Asia* 5: 1–3.

Flad, R., S. Li, X. Wu, and Z. Zhao. 2010. "Early wheat in China: Results from new studies at Donghuishan in the Hexi corridor." *The Holocene* 17: 555–560.

Fowler, Cary, and Pat Mooney. 1990. *Shattering: Food, Politics, and the Loss of Genetic Diversity*. Tucson: University of Arizona Press.

Frachetti, Michael D. 2012. "Multi-regional emergence of mobile pastoralism and non-uniform institutional complexity across Eurasia." *Current Anthropology* 53 (1): 2–38.

Frachetti, M.D., R.N. Spengler, G.J. Fritz, and A.N. Mar'yashev. 2010. "Earliest direct evidence for broomcorn millet and wheat in the Central Eurasian steppe region." *Antiquity* 84: 993–1010.

Fraser, James B. 1825. *Narrative of a Journey to Khorasan, in the Years 1821 and 1822*. London: Longman, Hurst, Rees, Orme, Brown, and Green.

Fritz, G.J., and B.D. Smith. 1988. "Old collections and new technology: Documenting the domestication of *Chenopodium* in eastern North America." *Midcontinental Journal of Archaeology* 13: 3–27.

Fu, D. 2001. "Discovery, identification and study of the remains of Neolithic cereals from the Changguogou site, Tibet." *Archaeology* 3: 66–74. (In Chinese.)

Fuchs, Jeff. 2008. "The Tea Horse Road." *Silk Road* 6 (1): 63–71.

Fuller, D.Q. 2001. "Responses: Harappan seeds and agriculture; Some considerations." *Antiquity* 75: 410–414.

Fuller, D. Q. 2002. "Fifty years of archaeobotanical studies in India: Laying a solid foundation." In *Indian Archaeology in Retrospect*, vol. 3, *Archaeology and Interactive Disciplines*, edited by S. Settar and R. Korisettar, 247–363. New Delhi: Manohar.

Fuller, D. Q. 2003. "African crops in prehistoric South Asia: A critical review." In *Food, Fuel and Fields: Progress in African Archaeobotany*, edited by K. Neumann, A. Butler, and S. Kahlheber, 239–271. Cologne: Heinrich-Barth Institute.

Fuller, D. Q. 2006. "Agricultural origins and frontiers in South Asia: A working synthesis." *Journal of World Prehistory* 20: 1–86.

Fuller, D. Q. 2008. "The spread of textile production and textile crops in India beyond the Harappan zone: An aspect of the emergence of craft specialization and systematic trade." In *Linguistics, Archaeology, and the Human Past*, edited by T. Osada and A. Uesugi, 1–26. Kyoto, Japan: Indus Project, Research Institute of Humanities and Nature.

Fuller, D. Q. 2011. "Finding plant domestication in the Indian subcontinent." *Current Anthropology* 52 (S4): S347–S362.

Fuller, D. Q. 2012. "Pathways to Asian civilizations: Tracing the origins and spread of rice and rice cultures." *Rice* 4: 9078.

Fuller, D. Q., and F. L. Zhang. 2007. "A preliminary report of the survey archaeobotany of the upper Ying Valley (Henan Pronince)." In *Archaeological Discovery and Research at the Wangchenggang Site in Elengfeng (2002–2005)*, 916–958. Zhengzhou: Great Elephant.

Fuller, D. Q., and N. Boivin. 2009. "Crops, cattle and commensals across the Indian Ocean: Current and potential archaeobiological evidence." *Études Océan Indien* 42–43: 13–46.

Fuller, D. Q., Emma Harvey, and Ling Qin. 2007. "Presumed domestication? Evidence for wild rice cultivation and domestication in the fifth millennium BC of the Lower Yangtze region." *Antiquity* 81: 316–331.

Fuller, D. Q., and M. Madella. 2001. "Issues in Harappan archaeobotany: Retrospect and prospect." In *Indian Archaeology in Retrospect*, vol. 2, *Protohistory*, edited by S. Settar and R. Korisettar, 317–390. New Delhi: Manohar.

Fuller, D. Q., and L. Qin. 2009. "Water management and labour in the origins and dispersal of Asian rice." *World Archaeology* 41: 88–111.

Fuller, Dorian Q., and Ling Qin. 2010. "Declining oaks, increasing artistry, and cultivating rice: The environmental and social context of the emergence of farming in the Lower Yangtze region." *Environmental Archaeology* 15 (2): 139–159.

Fuller, D. Q., and M. Rowlands. 2011. "Ingestion and food technologies: Maintaining differences over the long term in West, South and East Asia." In *Interweaving Worlds Systemic Interactions in Eurasia, 7th to 1st Millennium B.C.*, edited by T. Wilkinson, S. Sherratt, and J. Bennet, 37–60. Oxford: Oxbow.

Fuller, D. Q., L. Qin, Y. Zhang, Z. Zhao, X. Chen, L. Hosoya, and G. Sun. 2009. "The domestication process and domestication rate in rice: Spikelet bases from the Lower Yangtze." *Science* 323: 1607–1610.

Fuller, D. Q., T. Denham, M. Arroyo-Kalin, L. Lucas, C. J. Stevens, L. Qin, R. G. Allaby, and M. D. Purugganan. 2014. "Convergent evolution and parallelism in plant domestication revealed by an expanding archaeological record." *Proceedings of the National Academy of Sciences* 111 (17): 6147–6152.

Gegas, V. C., A. Nazari, S. Griffiths, J. Simmonds, L. Fish, S. Orford, L. Sayers, J. H. Doonan, and J. W. Snape. 2010. "A genetic framework for grain size and shape variation in wheat." *The Plant Cell* 22 (4): 1046–1056.

Ghosh, R., S. Gupta, S. Bera, H. E. Jiang, X. Li, and C. S. Li. 2008. "Ovi-caprid dung as an indicator of paleovegetation and paleoclimate in northwestern China." *Quaternary Research* 70: 149–157.

Golden, Peter B. 2011. *Central Asia in World History*. Oxford: Oxford University Press.

Gorbunova, N. G. 1986. *The Culture of Ancient Ferghana: VI Century B.C.–VI Century A.D.* Translated by A. P. Andryushkin. Vol. 281. Oxford: BAR International Series.

Gorbunova, Natalya. 1993. "Traditional movements of nomadic pastoralists and the role of seasonal migrations in the formation of ancient trade routes in Central Asia." *Silk Road Art and Archaeology* 3: 1–10.

Gordyagin, A. Ya. 1892. *Several Botanical Data on L. seeds*. Moscow: Daily General Physician's Office.

Grant, Mark. 2000. *Galen on Food and Diet*. London: Routledge.

Gross, Briana L., and Zhijun Zhao. 2014. "Archaeological and genetic insights into the origins of domesticated rice." *Proceedings of the National Academy of Sciences* 111 (17): 6190–6197.

Gupta, Anil K. 2004. "Origin of agriculture and domestication of plants and animals linked to early Holocene climate amelioration." *Current Science* 87 (1): 59.

Hamd-Allāh Mustawfī of Qazwīn. 1919 [1340]. *The Geographical Part of the* Nuzhat-Al-Qulūb. Translated by G. Le Strange. London: Luzac & Co.

Hanks, Bryan, and Kathrine Linduff, eds. 2009. *Social Complexity in Prehistoric Eurasia: Monuments, Metals and Mobility*. Cambridge: Cambridge University Press.

Hansen, Valerie. 2003. "The Hejia Village hoard: A snapshot of China's Silk Road trade." *Orientations* 34 (2): 14–19.

Hanson, Thor. 2015. *The Triumph of Seeds: How Grains, Nuts, Kernels, Pulses, and Pips Conquered the Plant Kingdom and Shaped Human History*. New York: Basic Books.

Harlan, J. R. 1971. "Agricultural origins: Centers and noncenters." *Science* 174: 468–474.

Harlan, J. R. 1975. *Crops and Man*. Madison, WI: American Society of Agronomy, Crop Science Society of America.

Harlan, J. R. 1977. "The origins of cereal agriculture in the Old World." In *Origins of Agriculture*, edited by Charles A. Reed, 357–383. Paris: Mouton.

Harlan, J. R., and D. Zohary. 1966. "Distribution of wild wheats and barley." *Science* 153: 1074–1080.

Harris, D. 2010. *Origins of Agriculture in Western Central Asia: An Environmental-Archaeological Study*. Philadelphia: University of Pennsylvania Museum of Archaeology and Anthropology.

Harris, S. A., Julian P. Robinson, and Barrie E. Juniper. 2002. "Genetic clues to the origin of the apple." *Trends in Genetics* 18: 426–430.

Harrison, N. 1995. "Preliminary archaeobotanical findings from Anau, 1994 excavations." *Harvard IuTAKE Excavations at Anau South, Turkmenistan* 28–36.

Helbaek, H. 1950. "Botanical study of the stomach contents of the Tollund man." *Aarbøger* 1: 329–341.

Helbaek, H. 1952. "Early crops in southern England." *Prehistoric Society*, 194–223. Copenhagen: National Museum, Copenhagen.

Helbaek, H. 1959. "Domestication of food plants in the Old World." *Science* 130: 365–372.

Helbaek, H. 1966. "The plant remains from Nimrud." In *Nimrud and Its Remains*, edited by M. E. L. Mallowan, 613–620. Edinburgh: Collins.

Herodotus. 1920 [ca. 450 BC]. *The Histories*. Translated by A.D. Godley. Cambridge, MA: Harvard University Press.

Herrmann, G., and K. Kurbansakhatov. 1994. "The international Merv project: Preliminary report on the second season (1993)." *Iran: British Institute of Persian Studies* 32: 53–75.

Hesiod. 1914 [eighth century BC]. *The Homeric Hymns and Homerica (Shield of Heracles)*. Translated by Hugh G. Evelyn-White. Cambridge, MA: Harvard University Press.

Hiebert, Fredrick T. 1994. *Origins of the Bronze Age Civilization of Central Asia*. Cambridge, MA: Peabody Museum of Archaeology and Ethnology, Harvard University.

Hiebert, Fredrick T. 2003. *A Central Asian Village at the Dawn of Civilization: Excavations at Anau, Turkmenistan*. Philadelphia: University of Pennsylvania Museum of Archaeology and Anthropology.

Higham, C. 1995. "The transition to rice cultivation in Southeast Asia." In *Last Hunters—First Farmers*, edited by T. Price and A. Gebauer, 127–155. Santa Fe, NM: School of American Research Press.

Hill, John E. 2009. *Through the Jade Gate to Rome: A Study of the Silk Routes during the Later Han Dynasty, 1st to 2nd Centuries CE: An Annotated Translation of the Chronicle on the "Western Regions" in the* Hou Hanshu. Lexington, KY: John E. Hill.

Ho, Ping-ti. 1975. *The Cradle of the East: An Inquiry into the Indigenous Origins of Techniques and Ideas of Neolithic and Early Historic China, 5000-1000 B.C.* Chicago: University of Chicago Press.

Huang, H.T. 2000. "Part 5: Fermentations and food science." In *Science and Civilisation in China, Volume VI: Biology and Biological Technology*. Cambridge: Cambridge University Press.

Huang, H., Z. Cheng, Z. Zhang, and Y. Wang. 2008. "History of cultivation and trends in China." In *The Peach: Botany, Production and Uses*, edited by D.R. Layne and D. Bassi, 37–60. Wallingford, CT: CABI.

Hudaikov, Y.S., S.G. Skobelev, O.A. Mitko, A.Y. Borisenko, and Zh. Orozbekova. 2013. "The burial rite of the early Scythian nomads of Tuva (based on the Bai-Dag I cemetery)." *Archaeology Ethnology and Anthropology of Eurasia* 41 (1): 104–113.

Humphrey, Caroline, Marina Mongush, and B. Telengid. 1994. "Attitudes to nature in Mongolia and Tuva: A preliminary report." In *Nomadic Peoples International Union of Anthropological and Ethnological Sciences Commission on Nomadic Peoples*, edited by Hjort af Ornäs Anders, 51–62. Montreal: Reprocentralen HSC.

Hunt, H. V., Xue Shang, and Martin K. Jones. 2017. "Buckwheat: A crop from outside the major Chinese domestication centres? A review of the archaeobotanical, palynological and genetic evidence." *Vegetation History and Archaeobotany* 27(3): 493–506.

Hunt, H. V., M. Vander Linden, X Liu, G. Motuzaite-Matuzeviciute, S. Colledge, and M. K. Jones. 2008. "Millets across Eurasia: Chronology and context of early records of the genera *Panicum* and *Setaria* from archaeological sites in the Old World." *Vegetation History and Archaeobotany* 17: S5–S18.

Hunt, H. V., M. G. Campana, M. C. Lawes, Y.-J. Park, M. A. Bower, C. J. Howe, and M. K. Jones. 2011. "Genetic diversity and phylogeography of broomcorn millet (*Panicum miliaceum* L.) across Eurasia." *Molecular Ecology* 20: 4756–4771.

Ibn al-Awwam. 2000 [twelfth century]. *Le livre de l'agriculture* (Book of agriculture). Translated by J. J. Clément-Mullet. Paris: Arles.

Ibn Fadlan. 2012. *Ibn Fadlan and the Land of Darkness Arab Travellers in the Far North*. Translated by Paul Lunde and Caroline Stone. London: Penguin Classics.

Ibn Hawqal. 1964. *Configuration de la terre (Kitab surat al-ard)*. Translated by Tome I. Kramers, J. H. and G. Wiet. Paris: Maisonneuve et Larose.

Jacomet, S. 2006. *Identification of Cereal Remains from Archaeological Sites*. Unpublished.

Jahangir. 1909–1914 [1569–1627]. *The Tūzuk-i-Jahangīrī: Memoirs of Jahāngīr (Tuzk-e-Jahangiri)*. Translated by Alexander Rogers and Henry Beveridge. London: Royal Asiatic Society.

Janik, L. 2002. "Wandering weed: The journey of buckwheat (*Fagopyrum* sp.) as an indicator of human movement in Russia." In *Ancient Interactions: East and West in Eurasia*, edited by K. Boyle, C. Renfrew, and M. Levine, 299–308. Cambridge: McDonald Institute for Archaeological Research.

Jashemski, W. F. 1979. *The Gardens of Pompeii: Herculaneum and the Villas Destroyed by Vesuvius*. New Rochelle, NY: Aristide de Caratzas.

Jia, P. W., A. Betts, and X Wu. 2011. "New evidence for Bronze Age agricultural settlements in the Zhunge'er (Junggar) Basin, China." *Journal of Field Archaeology* 36(4): 269–280.

Jiang, H. E., Y. B. Zhang, X. Li, Y. F. Yao, D. K. Ferguson, E. G. Lu, and C. S. Li. 2009. "Evidence for early viticulture in China: Proof of a grapevine (*Vitis vinifera* L., Vitaceae) in the Yanghai tombs, Xinjiang." *Journal of Archaeological Science* 36: 1458–1465.

Jiang, L., and L. Liu. 2006. "New evidence for the origins of sedentism and rice domestication in the Lower Yangzi River, China." *Antiquity* 80: 355–361.

Jing, Yuan, and Rod Campbell. 2009. "Recent archaeometric research on 'the origins of Chinese civilisation.'" *Antiquity* 83: 96–109.

Jones, G., and S. Valamoti. 2005. "Lallemantia, an imported or introduced oil plant in Bronze Age northern Greece." *Vegetation History and Archaeobotany* 14: 571–577.

Jones, G., H. Jones, M. P. Charles, M. K. Jones, S. Colledge, F. J. Leigh, D. A. Lister, L. M. J. Smith, W. Powell, and T. A. Brown. 2012. "Phylogeographic analysis of barley DNA as evidence for the spread of Neolithic agriculture through Europe." *Journal of Archaeological Science* 39 (10): 3230–3238.

Jones, G., M. P. Charles, M. K. Jones, S. Colledge, F. J. Leigh, D. A. Lister, and L. M. J. Smith. 2013. "DNA evidence for multiple introductions of barley into Europe following dispersed domestications in Western Asia." *Antiquity* 87: 701–713.

Jones, H., F. Leigh, I. Mackay, M. Bower, L. Smith, M. Charles, G. Jones, M. Jones, T. Brown, and W. Powell. 2008. "Population based resequencing reveals that the flowering time adaptation of cultivated barley originated east of the Fertile Crescent." *Molecular Biology and Evolution* 25: 2211–2219.

Jones, M. K., H. Hunt, E. Lightfoot, D. Lister, X. Liu, and G. Motuzaite-Matuzeviciute. 2011. "Food globalization in prehistory." *World Archaeology* 43 (4): 665–675.

Josekutty, P. C. 2008. "Defining the genetic and physiological basis of *Triticum Sphaerococcum Perc.*" MSc thesis, University of Canterbury.

Juniper, B. E., and D. J. Mabberley. 2006. *The Story of the Apple*. New York: Timber Press.

Karev, Y. 2004. "Samarqand in the eighth century: The evidence of transformation." In *Changing Social Identity with the Spread of Islam: Archaeological Perspectives*, edited by Donald Whitcomb, 51–66. Chicago: University of Chicago Press.

Karev, Y. 2013. "From Tents to Cities: The royal court of the western Qarakhanids between Bukhara and Samarkand." In *Turko-Mongol Rulers, Cities and City Life*, edited by David Durand-Guédy, 99–147. Boston: Brill.

Kawakami, S., K. Ebana, T. Nishikawa, Y. Sato, D. A. Vaughan, and K. Kadowaki. 2007. "Genetic variation in the chloroplast genome suggests multiple domestication of cultivated Asian rice (*Oryza sativa* L.)." *Genome* 50 (2): 180–187.

Kerje, T., and M. Grum. 2000. "The origin of melon, *Cucumis melo*: A review of the literature." *Acta Horticulturae* 510: 34–37.

Kim, M. 2013. "Wheat in ancient Korea: A size comparison of carbonized kernels." *Journal of Archaeological Science* 40: 517–525.

Kimata, M., E. G. Ashok, and A. Seetharam. 2000. "Domestication, cultivation and utilization of two small millets, *Brachiaria ramosa* and *Setaria glauca* (Poaceae), in South India." *Economic Botany* 54 (2): 217–227.

Kirleis, W., and E. Fischer. 2014. "Neolithic cultivation of tetraploid free threshing wheat in Denmark and Northern Germany: Implications for crop diversity and societal dynamics of the Funnel Beaker Culture." *Vegetation History and Archaeobotany* 23 (1): 81–96.

Kislev, M. E. 1985. "Early Neolithic horsebean from Yiftah'el, Israel." *Science* 228: 319–320.

Kislev, M. E. 1988. "Nahal Hemar cave: Desiccated plant remains; An interim report." *Atiqot* 18: 76–81.

Knörzer, K.-H. 2000. "3000 years of agriculture in a valley of the high Himalayas." *Vegetation History and Archaeobotany* 9: 219–222.

Koba, T., and K. Tsunewaki. 1978. "Mapping of the s and Ch 2 genes on chromosome 3D of common wheat." *Wheat Information Service* 45–46: 18–20.

Komatsuda, T., P. Mohammad, C. He, A. Perumal, K. Hiroyuki, D. Perovic, N. Stein, et al. 2007. "Six-rowed barley originated from a mutation in a homeodomain-leucine zipper I-class homeobox gene." *Proceedings of the National Academy of Sciences* 104 (4): 1424–1429.

Koroluyk, E. A., and N. V. Polosmak. 2010. "Plant remains from Moin Ula burial mounds 20 and 31 (Northern Mongolia)." *Archaeology, Ethnology and Anthropology of Eurasia* 38 (2): 57–63.

Koryakova, E. A., and A. V. Epimakhov. 2007. *The Urals and Western Siberia in the Bronze and Iron Ages*. Cambridge: Cambridge University Press.

Kovach, M. J., M. T. Sweeney, and S. R. McCouch. 2007. "New insights into the history of rice domestication." *Trends in Genetics* 11: 578–587.

Kradin, N. N. 2002. "Nomadism, evolution and world-systems: Pastoral societies in theories of historical development." *Journal of World-Systems Research* 8 (3): 368–388.

Kuzmina, E. 2008. *The Prehistory of the Silk Road: Encounters with Asia*. Translated by V. H. Mair. Philadelphia: University of Pennsylvania Press.

Landsell, Henry. 1885. *Russian Central Asia: Including Kuldja, Bokhara, Khiva, and Merv*. London: Sampson Low, Marston, Searle, and Rivington.

Langlie, BrieAnna S., Natalie G. Mueller, Robert N. Spengler, and Gayle J. Fritz. 2014. "Agricultural origins from the ground up: Archaeological perspectives on plant domestication." *American Journal of Botany* 101 (10): 1601–1617.

Lattimore, Owen. 1928. "Caravan routes of inner Asia." *Geographical Journal* 72 (6): 497–531.

Lattimore, Owen. 1940. *The Inner Asian Frontiers of China*. London: Beacon Press.

Lattimore, Owen. 1995 [1929]. *The Desert Road to Turkestan*. New York: Kodansha International.

Laufer, Berthold. 1919. *Sino-Iranica: Chinese Contributions to the History of Civilization in Ancient Iran, with Special Reference to the History of Cultivated Plants and Products*. Chicago: Field Museum of Natural History.

Leon, J. 2010. "Genetic diversity and population differentiation analysis of Ethiopian barley (*Hordeum vulgare* L.) landraces using morphological traits and SSR markers." Bonn: PhD diss., Rheinischen Friedrich-Wilhelms-Universität.

Levin, M. G., and L. P. Potapov. 1964. *The Peoples of Siberia*. Translated by Stephen Dunn. Chicago: University of Chicago Press.

Li, C., D. L. Lister, and H. Li. 2011. "Ancient DNA analysis of desiccated wheat grains excavated from a Bronze Age cemetary in Xinjiang." *Journal of Archaeological Science* 38: 115–119.

Li, H. L. 1969. "The vegetables of ancient China." *Economic Botany* 23: 253–260.

Li, Hui-Lin. 1959. *The Garden Flowers of China*. New York: Rowland.

Li, Hui-Lin. 1970. "The origins of domesticated plants in southeast Asia." *Economic Botany* 24: 3–19.

Li, S. 2002. "Interactions between northwest China and Central Asia during the second millennium BC: An archaeological perspective." In *Ancient Interactions: East and West in Eurasia*, edited by K. Boyle, C. Renfrew, and M. Levine, 171–182. London: McDonald.

Li, X., C. Xu, and Q. Zhang. 2004. "Ribosomal DNA spacer-length polymorphisms in three samples of wild and cultivated barleys and their relevance to the origin of cultivated barley." *Plant Breeding* 123: 30–34.

Li, X., J. Dodson, X. Zhou, H. Zhang, and R. Masutomoto. 2007. "Early cultivated wheat and broadening of agriculture in Neolithic China." *The Holocene* 15: 555–560.

Li, Y. 2007. "Animal bones and economy at the site of Karuo: An opinion on prehistoric agriculture in the Hengduan Mountain Chain." *Sichuan wenwu* 1: 50–56. (In Chinese.)

Li, Ya, Xiao Li, Hongyong Cao, Chunchang Li, Hongen Juang, and Chengsen Li. 2013. "Grain remains from archaeological sites and development of oasis agriculture in Turpan, Xinjiang." *Chinese Science* 58 (1): 40–45. (In Chinese.)

Lightfoot, Emma, Xinyi Liu, and Martin K. Jones. 2013. "Why move starchy cereals? A review of the isotopic evidence for prehistoric millet consumption across Eurasia." *World Archaeology* 45 (4): 574–623.

Linduff, K. M. 2006. "Why have Siberian artifacts been excavated within ancient Chinese dynastic borders?" In *Beyond the Steppe and the Sown: Proceedings of the 2002 University of Chicago Conference on Eurasian Archaeology*, edited by D. L. Peterson, L. M. Popova, and A. T. Smith, 358–370. Boston: Brill.

Lippolis, Carlo, and Niccoló Manassero. 2015. "Storehouses and storage practices in Old Nisa (Turkmenistan)." *Electrum* 22: 115–142.

Lisitsina, Gorislava N. 1984. "The Caucasus: A center of ancient farming in Eurasia." In *Plants and Ancient Man*, edited by W. van Zeist and W. A. Casparie, 285–292. Rotterdam: Balkema.

Lisitsyna, G. N., and L. V. Prishchepenko. 1977. *Palaeoethnobotanical finds of the Caucasus and the Near East*. Moscow: Nauka. (In Russian.)

Lister, D., and M. Jones. 2013. "Is naked barley an eastern or a western crop? The combined evidence of archaeobotany and genetics." *Vegetation History and Archaeobotany* 22 (5): 439–446.

Lister, D. L., S. Thaw, M. A. Bower, H. Jones, M. P. Charles, G. Jones, L. M. J. Smith, C. J. Howe, T. A. Brown, and M. K. Jones. 2009. "Latitudinal variation in a photoperiod response gene in European barley: Insight into the dynamics of agricultural spread from 'historic' specimens." *Journal of Archaeological Science* 38: 1090–1098.

Liu, C., Z. Kong, and S. D. Lang. 2004. "A discussion on agricultural and botanical remains and the human ecology of Dadiwan site." *Zhongyuan wenwu* (Cultural Relics of Central China) 4: 25–29. (In Chinese.)

Liu, Honggao, Yifu Cui, Xinxin Zuo, Haiming Li, Jian Wang, Dongju Zhang, Jiawu Zhang, and Guanghui Dong. 2016. "Human settlements and plant utilization since the late prehistoric period in the Nujiang River Valley, southeast Tibetan plateau." *Archaeological Research in Asia* 5: 63–71.

Liu, Li. 2004. *The Chinese Neolithic*. Cambridge: Cambridge University Press.

Liu, Li, Judith Field, Richard Fullagar, Chaohong Zhao, Xingcan Chen, and Jincheng Yu. 2010. "A functional analysis of grinding stones from an early Holocene site at Donghulin, North China." *Journal of Archaeological Science* 37: 2630–2639.

Liu, X., H. V. Hunt, and M. K. Jones. 2009. "River valleys and foothills: Changing archaeological perceptions of North China's earliest farms." *Antiquity* 83: 82–95.

Liu, Xinru. 2010. *The Silk Road in World History*. Oxford: Oxford University Press.

Liu, Xinyi, Diane L. Lister, Zhijun Zhao, Richard A. Staff, Penny Jones, Liping Zhou, Anubha Pathak, et al. 2016. "The virtues of small grain size: Potential pathways to a distinguishing feature of Asian wheats." *Quaternary International* 426: 107–119.

Livshits, V. A. 2015. "Sogdian epigraphy of Central Asia and Semirech'ye." In *Corpus Inscriptionum Iranicarum*, translated by Tom Stableford. London: School of Oriental and African Studies.

Lone, F. A., M. Khan, and G. M. Buth. 1993. *Palaeoethnobotany Plants and Ancient Man in Kashmir*. New Delhi: Oxford and IBH.

Lu, Hongliang. 2016. "Colonization of the Tibetan plateau, permanent settlement, and the spread of agriculture: Reflection on current debates on the prehistoric archeology of the Tibetan plateau." *Archaeological Research in Asia* 5: 12–15.

Lu, Houyuan, Jianping Zhang, Kam-biu Liu, Naiqin Wu, Yumei Li, Kunshu Zhou, Maolin Ye, et al. 2009a. "Earliest domestication of common millet (*Panicum miliaceum*) in East Asia extended to 10,000 years ago." *Proceedings of the National Academy of Sciences* 106: 7367–7372.

Lu, Houyan, J. Zhang, N. Wu, K-b. Liu, D. Xu, and Q. Li. 2009b. "Phytolith analysis for the discrimination of foxtail millet (*Setaria italica*) and common millet (*Panicum miliaceum*)." *PLoS ONE* 4(2): e4448.

Lu, Houyuan, Jianping Zhang, Yimin Yang, Xiaoyan Yang, Baiqing Xu, Wuzhan Yang, Tao Tong, et al. 2016. "Earliest tea as evidence for one branch of the Silk Road across the Tibetan plateau." *Science Reports* 6: 18955.

Lunina, S. B. 1984. *Cities in Southern Sogda in the seventh to tenth centuries*. Tashkent: FAN. (In Russian.)

Lyre, P. B. 2012. *Material from the Penjikent Archaeological Expedition*. Vol. 14. St. Petersburg: Government Heritage. (In Russian.)

Ma, D., T. Xu, M. Gu, B. Wu, and Y. Kang. 1987. "The classification and distribution of wild barley in the Tibet autonomous region." *Scientia Agriculturae Sinica* 20: 1–6.

MacGahan, J. A. 1876. *Campaigning on the Oxus and the Fall of Khiva*. London: Sampson Low, Maeston, Searle, & Rivington.

Mair, Victor H., and Erling Hoh. 2009. *The True History of Tea*. London: Thames and Hudson.

Maksudov, Farhad, Elissa Bullion, Edward R. Henry, Taylor Hermes, Ann M. Merkle, and Michael D. Frachetti. In press. "Nomadic Urbanism at Tashbulak: A New Highland Town of the Qarakhanids." In *Proceedings of the Urban Culture of Central Asia Conference, Society for the Exploration of Eurasia*. Bern, Switzerland.

Mani, B. R. 2008. "Kashmir Neolithic and early Harappan: A linkage." *Pragdhara* 18: 229–247.

Marinova, E. and S. Riehl. 2009. "*Carthamus* species in the ancient Near East and south-eastern Europe: Archaeobotanical evidence for their distribution and use as a source of oil." *Vegetation History and Archaeobotany* 18: 341–349.

McGovern, P. E. 2003. *Ancient Wine: The Search for the Origins of Viniculture*. Princeton, NJ: Princeton University Press.

McGovern, P. E., D. L. Glusker, L. J. Exner, and M. M. Voigt. 1996. "Neolithic resinated wine." *Nature* 381: 480–481.

McNally, K. L., K. L. Childs, R. Bohnert, R. M. Davidson, K. Zhao, V. J. Ulat, G. Zeller, et al. 2009. "Genomewide SNP variation reveals relationships among landraces and modern varieties of rice." *Proceedings of the National Academy of Sciences* 106 (30): 12273–12278.

Miller, N. 1981. "Plant remains from Villa Royale II, Susa." *Cahiers de la Délégation Archéologique Française en Iran* 12: 37–42.

Miller, N. F. 1990. "Godin Tepe, Iran: Plant remains from period V, the late fourth millennium B.C." *Museum Applied Science Center for Archaeology Ethnobotanical Report* 6: 1–12.

Miller, N. F. 1999. "Agricultural development in western Central Asia in the Chalcolithic and Bronze Ages." *Vegetation History and Archaeobotany* 8: 13–19.

Miller, N. F. 2003. "The use of plants at Anau North." In *A Central Asian Village at the Dawn of Civilization: Excavations at Anau, Turkmenistan*, edited by F. T. Hiebert and K. Kurdansakhatov. Philadelphia: University of Pennsylvania Museum.

Miller, N. 2008. "Sweeter than wine? The use of the grape in early western Asia." *Antiquity* 82: 937–946.

Miller, N. F. 2010. *Botanical Aspects of Environment and Economy at Gordion, Turkey*. Philadelphia: University of Pennsylvania Museum.

Miller, N. F. 2011. "Preliminary archaeobotanical results." In *Excavations at Monjukli Depe, Meana-Čaača Region, Turkmenistan, 2010*, edited by S. Pollock and R. Bernbeck, 43: 213–221. Berlin: Archäologische Mitteilungen aus Iran und Turan.

Miller, Naomi F., Robert N. Spengler III, and Michael D. Frachetti. 2016. "Millet cultivation across Eurasia: Origins, spread, and the influence of seasonal climate." *The Holocene* 26: 1566–1575.

Millward, James A. 2013. *The Silk Road: A Very Short Introduction*. Oxford: Oxford University Press.

Milton, John. 2004 [1667]. *Paradise Lost*. Edited by David Hawker. New York: Barnes & Noble Classics.

Miquel, A. 1980. *La géographie humaine du monde musulman jusqu'au milieu du IIe siècle*, volume 3, *Le milieu naturel*. Paris: Mouton.

Mirsky, Jeannette. 1977. *Sir Aurel Stein: Archaeological Explorer.* Chicago: University of Chicago Press.

Molina-Cano, J., M. Moralejo, E. Igartua, and I. Romagosa. 1999. "Further evidence supporting Morocco as a center of origin of barley." *Theoretical and Applied Genetics* 98: 913–918.

Molina-Cano, J., J. Russell, M. Moralejo, J. Escacena, G. Arias, and W. Powell. 2005. "Chloroplast DNA microsatellite analysis supports a polyphyletic origin for barley." *Theoretical and Applied Genetics* 110 (4): 613–619.

Moore, K., N.F. Miller, F.T. Heibert, and R.H. Meadow. 1994. "Agriculture and herding in early oasis settlements of the Oxus civilization." *Antiquity* 68: 418–427.

Morgan, C., L. Barton, R.L. Bettinger, and F. Chen. 2011. "Glacial cycles and Palaeolithic adaptive variability on China's western loess plateau." *Antiquity* 85: 365–379.

Morrell, P.L., and M.T. Clegg. 2007. "Genetic evidence for a second domestication of barley (*Hordeum vulgare*) east of the Fertile Crescent." *Proceedings of the National Academy of Sciences* 104: 3289–3294.

Morrell, P.L., K.E. Lundy, and M.T. Clegg. 2003. "Distinct geographic patterns of genetic diversity are maintained in wild barley (*Hordeum vulgare* ssp. *spontaneum*) despite migration." *Proceedings of the National Academy of Sciences* 100: 10812–10817.

Motuzaite-Matuzeviviute, G., S. Telizhenko, and M.K. Jones. 2012. "Archaeobotanical investigation of two Scythian-Sarmatian period pits in eastern Ukraine: Implications for floodplain cereal cultivation." *Journal of Field Archaeology* 37: 51–61.

Motuzaite-Matuzeviciute, G., R.A. Staff, H.V. Hunt, X. Liu, and M.K. Jones. 2013. "The early chronology of broomcorn millet (*Panicum miliaceum*) in Europe." *Antiquity* 87: 1073–1085.

Motuzaite-Matuzeviciute, G., E. Lightfoot, T.C. O'Connell, D. Voyakin, X. Liu, V. Loman, S. Svyatko, E. Usmanova, and M.K. Jones. 2015. "The extent of cereal cultivation among the Bronze Age to Turkic period societies of Kazakhstan determined using stable isotope analysis of bone collagen." *Journal of Archaeological Science* 59: 23–34.

Motuzaite Matuzeviciute, G., R.C. Preece, S. Wang, L. Colominas, K. Ohnuma, S. Kume, A. Abdykanova, and M.K. Jones. 2016. "Ecology and subsistence at the Mesolithic and Bronze Age site of Aigyrzhal-2, Naryn valley, Kyrgyzstan." *Quaternary International* 437: 35–49.

Mowart, Farley. 1970. *The Siberians*. New York: Little, Brown and Company.

Murdock, G. P. 1959. *Africa: Its Peoples and Their Culture History*. New York: McGraw-Hill.

Murphy, C. 2016. "Finding millet in the Roman world." *Archaeological and Anthropological Sciences* 8 (1): 65–78.

Murphy, C., G. Thompson, and D. Q. Fuller. 2013. "Roman food refuse: Urban archaeobotany in Pompeii, regio VI, insula I." *Vegetation History and Archaeobotany* 22: 409–419.

Murphy, Eileen M., Rick Schulting, Nick Beer, Yuri Chistov, Alexey Kasparov, and Margarita Psenitsyna. 2013. "Iron Age pastoral nomadism and agriculture in the eastern Eurasian steppe: Implications from dental paleopathology and stable carbon and nitrogen isotopes." *Journal of Archaeological Science* 40: 2547–2560.

Myles, Sean, Adam R. Boyko, Christopher L. Owens, Patrick J. Brown, Fabrizio Grassi, Mallikarjuna K. Aradhya, Bernard Prins, et al. 2010. "Genetic structure and domestication history of the grape." *Proceedings of the National Academy of Sciences* 108 (9): 3530–3535.

Nabhan, Gary Paul. 2014. *Cumin, Camels, and Caravans: A Spice Odyssey*. Berkeley: University of California Press.

Neef, R. 1989. "Plants." In *Picking Up the Threads: A Continuing Review of Excavations at Deir Alla, Jordan*. Edited by G. Van der Kooij and M. M. Ibrahim, 30–37. Leiden: Archaeological Centre, University of Leiden.

Nesbitt, M. 1994. "Archaeobotanical research in the Merv Oasis." In *The International Merv Project, Preliminary Report on the Second Season*, edited by K. Kurbansakhatov and G. Herrmann, 53–75. Unpublished field report.

Nesbitt, M. R., and G. Summers. 1988. "Some recent discoveries of millet (*Panicum miliaceum* L. and *Setaria italica* (L.) P. Beauv.) at excavations in Turkey and Iran." *Anatolian Studies* 38: 85–97.

Nesbitt, Mark, St John Simpson, and Ingvar Svanberg. 2010. "History of rice in western and central Asia." In *Rice: Origin, Antiquity and History*, edited by S. D. Sharma, 308–340. London: Science Publishers.

Norwich, John Julius. 1989. *Byzantium (I): The Early Centuries*. New York: Knopf.

Nout, M. J. R., G. Tuncle, and L. Brimer. 1995. "Microbial degradation of amygdalin of bitter apricot seeds (*Prunus armeniaca*)." *International Journal of Food Microbiology* 24: 407–412.

O'Donovan, Edmond. 1883. *Merv: A story of adventures and captivity epitomized from "the Merv Oasis."* London: Smith, Elder.

Ohnishi, O. 1998. "Search for the wild ancestor of buckwheat: The wild ancestor of cultivated common buckwheat, and of tatary buckwheat." *Economic Botany* 52: 123–133.

Ohnishi, Ohmi. 2004. "On the origin of cultivated buckwheat." *Proceedings of the 9th International Symposium on Buckwheat, Prague 2004* 1: 16–21.

Olearius, Adam. 2004 [1603–1671]. In *The Voyages and Travels of the Ambassadors Sent by Frederick, Duke of Holstein, to the Great Duke of Muscovy and the King of Persia*. Translated by John Davies. Ann Arbor, MI: Text Creation Partnership.

Osbaldeston, Tess Anne. 2000. "Introduction." In *Dioscorides: De Materia Medica, by Pedanius Dioscorides*. Johannesburg: Ibidis Press.

Pashkevich, G. 1984. "Palaeoethnobotanical examination of archaeological sites in the Lower Dnieper region, dated to the last centuries BC and first centuries AD." In *Plants and Ancient Man: Studies in Palaeoethnobotany*, edited by W. van Zeist and W. A. Casparie, 277–284. Boston: A. A. Balkema.

Paskhevich, G. 2003. "Palaeoethnobotanical examination of archaeological sites in the steppe and forest-steppe of East Europe in the Late Neolithic and Bronze Age." In *Prehistoric Steppe Adaptation and the Horse*, edited by M. Levine, C. Renfrew, and K. Boyle, 287–297. Cambridge: McDonald Institute.

Pellat, Charles. 1954. "Ğāhiziana I: Le Kitāb al-Tabassur bi-l-Tiğāra Attribué á Ğāhiz (Gahiziana I: The book al-Tabari attributed to al-Jahiz)." *Arabica: Revue d'Études Arabes* 1 (2): 153–165.

Percival, J. 1921. *The Wheat Plant*. London: Duckworth.

Perry, Charles. 2017 [thirteenth century]. In *Scents and Flavors the Banqueter Favors (Kitab al-Wuslah ila l-Habib fı Wasf al-Tayyibat wal-Tib)*. Translated by Charles Perry. New York: New York University Press.

Peterson, R. F. 1965. *Wheat: Botany, Cultivation, and Utilization*. New York: Leonard Hill/Interscience Publishers.

Pilipko, V. N. 2001. *Old Nisa: Primary Results of Excavations during the Soviet Period*. Moscow: Nauka. (In Russian.)

Pliny the Elder. 1855 [AD 77–79]. *Naturalis Historia* (Natural history). Translated by John Bostock. London: Taylor and Francis.

Pollan, Michael. 2001. *The Botany of Desire*. New York: Random House.

Pollegioni, Paola, Keith E. Woeste, Francesca Chiocchini, Stefano Del Lungo, Irene Olimpieri, Virginia Tortolano, Jo Clark, Gabriel E. Hemery, Sergio Mapelli, and Maria Emilia Malvolti. 2017. "Ancient humans influenced the current spatial genetic structure of common walnut populations in Asia." *PLoS ONE* 10 (9): e0135980.

Pollock, Susan. 2015. "Ovens, fireplaces and the preparation of food in Uruk Mesopotamia." *Origini: Preistoria e Protostoria Delle Civilta Antiche/Prehistory and Protohistory of Ancient Civilizations* 37 (1): 35–37.

Polo, Marco. 1845 [ca. 1300]. In *The Travels of Marco Polo*, edited by Hugh Murray. New York: Harper.

Polybius. 1962 [ca. 140 BC]. *Histories*. Translated by Evelyn S. Shuckburgh. New York: Bloomington.

Popov, A. A. 1966. *The Nganasan: The Material Culture of the Tavgi Samoyeds*. Translated by Elaine K. Ristinen. Indianapolis: Indiana University Press.

Popov, H. P. 1803. "Hungry bread, etc." *Medical Review* 12: 1803. (In Russian.)

Popova, L. M. 2006. "Political pastures: Navigating the steppe in the middle Volga REGION (Russia) during the Bronze Age." PhD diss., University of Chicago.

Pourkheirandish, M., and T. Komatsuda. 2007. "The importance of barley genetics and domestication in a global perspective." *Annual Review of Botany* 100 (5): 999–1008.

Priklonskii, V. L. 1953 [1881]. *Three Years in the Yakut Territory (Ethnographic Sketches): Yakut Ethnographic Sketches, Parts 1 and 2*. Translated by Sheldon Wise. New Haven, CT: Human Relations Area Files.

Qian, S. 1993 [91–109 BC]. *Records of the Great Historian: Han Dynasty and Qin Dynasty*. Volume 3. Translated by B. Watson. New York: Columbia University Press.

Quinn, Elizabeth A., Kesang Diki Bista, and Geoff Childs. 2015. "Milk at altitude: Human milk macronutrient composition in a high-altitude adapted population of Tibetans." *American Journal of Physical Anthropology* 159 (2): 233–243.

Ranov, V., and M. Bubnova. 1961. "Uncovering the history of the roof of the world." *American Journal of Archaeology* 65 (4): 396–397.

Rao, M. V. P. 1977. "Mapping of the sphaerococcum gene 's' on chromosome 3D of wheat." *Cereal Research Communication* 5: 15–17.

Renfrew, Colin. 2014. "Foreword: The Silk Road before silk." In *Reconfiguring the Silk Road: New Research on East-West Exchange in Antiquity*, edited by Victor H. Mair and Jane Hickman, xi–xiv. Philadelphia: University of Pennsylvania Museum of Archaeology and Anthropology.

Renfrew, J. M. 1987. "Fruits from ancient Iraq: The paleoethnobotanical evidence." *Bulletin of Sumerian Agriculture* 3: 157–161.

Reynolds, M. P., and N. E. Borlaug. 2006. "Impacts of breeding on international collaborative wheat improvement." *Journal of Agricultural Science* 144: 3–17.

Rhindos, David. 1984. *The Origins of Agriculture: An Evolutionary Perspective.* Waltham, MA: Academic Press.

Rhode, David, David B. Madsen, Jeffery P. Brantingham, and Tsultrim Dargye. 2007. "Yaks, yak dung, and prehistoric human habitation of the Tibetan plateau." *Developments in Quaternary Sciences* 9: 205–224.

Richthofen, Ferdinand von. 1877. "Über die zentralasiatischen Seidenstrassen bis zum 2. Jh. n. Chr. (On the Central Asian Silk Roads until the 2nd century AD)." *Verhandlungen der Gesellschaft für Erdkunde zu Berlin* (Proceedings of the Society for Geography in Berlin) 1877: 96–122.

Riehl, S. 1999. *Bronze Age Environment and Economy in the Troad: The Archaeobotany of Kumtepe and Troy; BioArchaeologica 2.* Tübingen: Mo-Vince-Verlag.

Riehl, S., M. Zeidi, and N. J. Conard. 2013. "Emergence of agriculture in the foothills of the Zagros mountains of Iran." *Science* 341 (6141): 65–67.

Rogers, J. D. 2007. "The contingencies of state formation in eastern Inner Asia." *Asian Perspectives* 46 (2): 249–274.

Rosen, A. M., C. Chang, and F. P. Grigoriev. 2000. "Paleoenvironments and economy of Iron Age Saka-Wusun agro-pastoralists in southeastern Kazakhstan." *Antiquity* 74: 611–623.

Rouse, L., and B. Cerasetti. 2014. "Ojakly: A late Bronze Age mobile pastoral site in the Murghab, Turkmenistan." *Journal of Field Archaeology* 39 (1): 32–50.

Salina, E., A. Borner, I. Leonova, V. Korzun, L. Laikova, O. Maystrenko, and S. Röder. 2000. "Microsatellite mapping of the induced sphaerococcoid mutation genes in *Triticum aestivum.*" *Theoretical and Applied Genetics* 100: 686–689.

Salunkhe, A., S. Tamhankar, S. Tetali, M. Zaharieva, D. Bonnett, R. Trethowan, and S. Misra. 2012. "Molecular genetic diversity analysis in emmer wheat (*Triticum dicoccon* Schrank) from India." *Genetic Resources and Crop Evolution* 60 (1): 165–174.

Salvi, Sergio. 2013. "Ipotesi sulle origini del 'grano di Rieti' (Hypothesis on the origins of the 'grain of Rieti')." *Proposte e ricerche* 36 (71): 233–238.

Salvi, Sergio. 2014. "Alle origini del concetto di prodotto tipico: Il caso del grano di Rieti (The origins of the concept of the typical product: The case of Rieti grain)." *Proposte e ricerche* 37 (73): 205–208.

Samuel, Delwen. 2001. "Archaeobotanical evidence and analysis." In *Mission Mésopotamie syrienne: Archéologie islamique (1986–1989); Peuplement rural et aménagments hydroagricoles dans la moyenne vallée de l'Eurphrate, fin VIIe–XIXe Siècle*, edited by Sophie Berthier, Louis Chaix, Jacqueline Studer, Oliver D'Hont, Rike Gyselend, and Delwen Samuel, 347–481. Damascus, Syria: Institut Français de Damas.

Sang, T., and S. Ge. 2007a. "The puzzle of rice domestication." *Journal of Integrated Plant Biology* 49 (6): 760–768.

Sang, T., and S. Ge. 2007b. "Genetics and phylogenetics of rice domestication." *Current Opinion in Genetics and Development* 17 (6): 533–538.

Saraswat, K. S. 1986. "Ancient crop plant remains from Springverpura, Allahabad (c. 1050–1000 BC)." *Geophytology* 16: 97–106.

Saraswat, K. S., and A. K. Pokharia. 2003. "Palaeoethnobotanical investigations at Early Harappan Kunal." *Pragdhara* 12: 105–140.

Saraswat, K. S., and A. K. Pokharia. 2004. *Plant Resources in the Neolithic Economy at Kanishpur, Kashmir.* Lucknow, India: Joint Annual Conference of the Indian Archaeological Society, Indian Society of Prehistoric and Quaternary Studies, Indian History and Culture Society.

Scarborough, John. 1978. "Theophrastus on herbals and herbal remedies." *Journal of the History of Biology* 11 (2): 353–385.

Schafer, E. H. 1963. *The Golden Peaches of Samarkand*. Berkeley: University of California Press.

Schuyler, Eugene. 1877. *Turkistan: Notes of a Journey in Russian Turkistan, Khokand, Bukhara, and Kuldja*. Volume 1. New York: Scribner, Armstrong & Co.

Sebastian, Patrizia, Hanno Schaefer, Ian R. H. Telford, and Susanne S. Renner. 2010. "Cucumber (*Cucumis sativus*) and melon (*C. melo*) have numerous wild relatives in Asia and Australia, and the sister species of melon is from Australia." *Proceedings of the National Academy of Sciences* 107 (32): 14269–14273.

Seebohm, Henry. 1882. *Siberia in Asia: A Visit to the Valley of the Yenesay in East Siberia*. London: William Clowes and Sons.

Selens, Ahmed, and Michael Freeman. 2011. "Pu-erh tea and the southwest Silk Road: An ancient quest for well-being." *Herbal Gram* 90: 32–43.

Semple, Ellen Churchill. 1928. "Ancient Mediterranean agriculture, part I." *Agricultural History* 2 (61–98): 2.

Serventi, Silvano, and Françoise Sabban. 2002. *Pasta: The Story of a Universal Food*. Translated by Antony Shugaar. New York: Columbia University Press.

Shantz, H. L., and L. N. Piemeisel. 1927. "The water requirement of plants at Akron, Colo." *Journal of Agricultural Research* 34: 1093–1190.

Sharma, A. K. 2000. *Early Man in Jammu Kashmir and Ladakh*. New Delhi: Agam Kala Prakashan.

Shaw, F. J. P. 1943. "Vegetation remains." In *Chanhu-Daro Excavations, 1935–36*, edited by E. J. H. Mackay, 250–251. New Haven, CT: Yale University Press.

Sherratt, A. G. 1981. "Plough and pastoralism: Aspects of the secondary products revolution." In *Patterns of the Past: Studies in Honour of David Clark*, edited by I. Hodder, G. Isaac, and N. Hammond, 261–305. Cambridge: Cambridge University Press.

Sherratt, A. G. 1983. "The secondary exploitation of animals in the Old World." *World Archaeology* 15: 90–104.

Shishlina, N. 2008. *Reconstruction of the Bronze Age of the Caspian Steppe: Life Styles and Life Ways of Pastoral Nomads*. Oxford: BAR International Series 1876.

Siculus, Diodorus. 1967 [ca. 60 BC]. *Diodorus of Sicily in Twelve Volumes, Book II*. Translated by C. H. Oldfather. London: Wlliam Heineman Ltd.

Silva, Fabio, Chris Stevens, Alison Weisskopf, Cristina Castillo, Ling Qin, Andrew Bevan, and Dorian Q. Fuller. 2015. "Modelling the origin of rice farming in Asia using the Rice Archaeological Database." *PloS ONE* 10 (9): e0137024.

Simonson, Tatum S., Yingzhong Yang, Chad D. Huff, Haixia Yun, Ga Qin, David J. Witherspoon, Zhenzhong Bai, et al. 2010. "Genetic evidence for high-altitude adaptation in Tibet." *Science* 329 (5987): 72–75.

Simoons, Fredrick J. 1990. *Food in China: A Cultural and Historical Inquiry.* Boca Raton, FL: CRC Press.

Singh, O. P. 2008. "Indian Ocean dipole mode and tropical cyclone frequency." *Current Science* 94 (1): 29–31.

Singh, R. 1946. "*Triticum sphaerococcum* Perc. (Indian dwarf wheat)." *Indian Journal of Genetics* 6: 34–47.

Snir, Ainit, and Ehud Weiss. 2014. "A novel morphometric method for differentiating wild and domesticated barley through intra-rachis measurements." *Journal of Archaeological Science* 44: 69–75.

Song, Jixiang, Hongliang Lu, Zhengwei Zhang, and Xinyi Liu. 2018. "Archaeobotanical remains from the mid-first millennium AD site of Kaerdong in western Tibet." *Archaeological and Anthropological Sciences.* Online first.

Soucek, S. 2000. *A History of Inner Asia.* Cambridge: Cambridge University Press.

Spengler, R. N. 2013. "Botanical resource use in the Bronze and Iron Age of the Central Eurasian mountain/steppe interface: Decision making in multiresource pastoral economies." PhD diss., Washington University in St. Louis.

Spengler III, R. N. 2014. "Niche dwelling vs. niche construction: Landscape modification in the Bronze and Iron Ages of Central Asia." *Human Ecology* 42 (6): 813–821.

Spengler III, R. N. 2015. "Agriculture in the Central Asian Bronze Age." *Journal of World Prehistory* 28 (3): 215–253.

Spengler III, R. N., and G. Willcox. 2013. "Archaeobotanical results from Sarazm, Tajikistan, an Early Bronze Age village on the edge: Agriculture and exchange." *Journal of Environmental Archaeology* 18 (3): 211–221.

Spengler III, R. N., C. Chang, and P. A. Tortellotte. 2013. "Agricultural production in the Central Asian mountains, Tuzusai, Kazakhstan." *Journal of Field Archaeology* 38(1): 68–85.

Spengler, Robert N., Michael D. Frachetti, and Gayle J. Fritz. 2013. "Ecotopes and herd foraging practices in the steppe/mountain ecotone of Central Asia during the Bronze and Iron Age." *Journal of Ethnobiology* 33 (1): 125–147.

Spengler III, R. N., P. N. Doumani, and M. D. Frachetti. 2014. "Late Bronze Age agriculture at Tasbas in the Dzhungar Mountains of eastern Kazakhstan." *Quaternary International* 348: 147–157.

Spengler III, R. N., B. Cerassetti, M. Tengberg, M. Cattani, and L. M. Rouse. 2014a. "Agriculturalists and pastoralists: Bronze Age economy of the Murghab Delta, southern central Asia." *Journal of Vegetation History and Archaeobotany* 23: 805–820.

Spengler III, R. N., M. D. Frachetti, P. N. Doumani, L. M. Rouse, B. Cerasetti, E. Bullion, and N. Mar'yashev. 2014b. "Early agriculture and crop transmission among Bronze Age mobile pastoralists of Central Eurasia." *Proceedings of the Royal Society B* 281: 2013.3382.

Spengler III, Robert N., Natalia Ryabogina, Pavel Tarasov, and Mayke Wagner. 2016. "The spread of agriculture into northern Central Asia." *The Holocene* 26: 1523–1526.

Spengler III, Robert N., Naomi F. Miller, Reinder Neef, Perry A. Tourtellotte, and Claudia Chang. 2017a. "Linking agriculture and exchange to social developments of the Central Asian Iron Age." *Journal of Athropological Archaeology* 48: 295–308.

Spengler III, Robert N., Ilaria de Nigris, Barbara Cerasetti, and Lynne M. Rouse. 2017b. "The breadth of dietary economy in the Central Asian Bronze Age: A case study from Adji Kui in the Murghab Region of Turkmenistan." *Journal of Archaeological Science*. Online first.

Spengler III, Robert N., Farhod Maksudov, Elissa Bullion, Ann Merkle, Taylor Hermes, and Michael D. Frachetti. 2018. "Arboreal crops on the Medieval Silk Road: Archaeobotanical studies at Tashbulak." *PLoS ONE* 13(8): e0201409

Stapf, O. 1931. "Comments on cereal and fruits." In *Mohenjodaro and the Indus Civilization*, edited by J. Marshall. London: Arthur Probsthain.

Stefanovsky, F. K. 1893. Materials for studying the properties of hungry bread. Moscow. (In Russian.)

Stein, Aurel. 1907. *Ancient Khotan*. Volume 1. Oxford: Oxford University Press.

Stein, Aurel. 1998 [1932]. *On Ancient Central Asian Tracks*. New Delhi: South Asia Books.

Sterckx, Roel. 2011. *Food, Sacrifice, and Sagehood in Early China*. Cambridge: Cambridge University Press.

Stevens, Chris J., Charlene Murphy, Rebecca Roberts, Leilani Lucas, Fabio Silva, and Dorian Q. Fuller. 2016. "Between China and South Asia: A Middle Asian corridor of crop dispersal and agricultural innovation in the Bronze Age." *The Holocene* 26 (10): 1541–1555.

Strabo. 1924 [7 BC–AD 23]. *The Geography of Strabo*. Translated by H.L. Jones. Cambridge, MA: Harvard University Press.

Su, Tao, Peter Wilf, Yongjiang Huang, Shitao Zhang, and Zhekun Zhou. 2015. "Peaches preceded humans: Fossil evidence from SW China." *Scientific Reports* 5: 16794.

Subtelny, Maria Eva. 2013. "Agriculture and the Timurid Chahārbāgh: The evidence from a medieval Persian agricultural manual." In *Turko-Mongol Rulers, Cities and City Life*, edited by David Durand-Guédy, 110–128. Boston: Brill.

Sumner, William M., and Donald Whitcomb. 1999. "Islamic settlement and chronology in Fars: An archaeological perspective." *Iranica Antiqua* 34: 309–324.

Svyatko, Svetlana V., Rick J. Schulting, James Mallory, Eileen M. Murphy, Paula J. Reimer, Valeriy I. Khartanovich, Yury K. Chistov, and Mikhail V. Sablin. 2013. "Stable isotope dietary analysis of prehistoric populations from the Minusinsk Basin, southern Siberia, Russia: A new chronological framework for the introduction of millet to the eastern Eurasian Steppe." *Journal of Archaeological Sciences* 40 (11): 3936–3945.

Tafuri, M.A., O.E. Graig, and A. Canci. 2009. "Staple isotope evidence for the consumption of millet and other plants in Bronze Age Italy." *American Journal of Physical Anthropology* 139: 146–153.

Takahashi, R. 1972. "Non-brittle rachis 1 and non-brittle rachis 2." *Barley Genetics Newsletter* 2: 181–182.

Takahashi, R., S. Hayashi, S. Yasuda, and U. Hiura. 1963. "Characteristics of the wild and cultivated barleys from Afghanistan and its neighbouring regions." *Bericht des Ohara Instituts für Landwirtschaftliche Biologie, Okayama* 1: 1–23.

Takahashi, R., S. Hayashi, U. Hiura, and S. Yasuda. 1968. "A study of cultivated barleys from Nepal, Himalaya and North India, with special reference to their phylogenetic differentiation." *Bericht des Ohara Instituts für Landwirtschaftliche Biologie* 14: 85–122.

Taketa, Shin, Satoko Amano, Yasuhiro Tsujino, Tomohiko Sato, Daisuke Saisho, Katsuyuki Kakeda, Mika Nomura, et al. 2008. "Barley grain with adhering hulls is controlled by an ERF family transcription factor gene regulating a lipid biosynthesis pathway." *Proceedings of the National Academy of Sciences* 105 (10): 4062–4067.

Tang, H., G. J. Wyckoff, J. Lu, and C. I. Wu. 2004. "A universal evolutionary index for amino acid changes." *Molecular Biology and Evolution* 21: 1548–1556.

Tanno, K., and G. Willcox. 2006a. "How fast was wild wheat domesticated?" *Science* 311 (1): 886.

Tanno, K., and G. Willcox. 2006b. "The origins of cultivation of *Cicer arietinum* L. and *Vicia faba* L.: Early finds from Tell el-Kerkh, north-west Syria, late 10th millennium BP." *Vegetation History and Archaeobotany* 15 (3): 197–204.

Tanno, K., and G. Willcox. 2012. "Distinguishing wild and domestic wheat and barley spikelets from early Holocene sites in the Near East." *Vegetation History and Archaeobotany* 21 (2): 107–115.

Tanno, K., and K. Takeda. 2004. "On the origin of six-rowed barley with brittle rachis, agriocrithon [*Hordeum vulgare* ssp. *vulgare f. agriocrithon* (Aberg) Bowd.], based on a DNA marker closely linked to the vrs1 (six-row gene) locus." *Theoretical Applications in Genetics* 110 (1): 145–150.

Tengberg, M. 1999. "Crop husbandry at Miri Qalat, Makran, SW Pakistan (4000–2000 BC)." *Vegetation History and Archaeobotany* 8 (1–2): 3–12.

Theophrastus. 1916 [ca. 350–287 BC]. *Enquiry into Plants, Books I–V.* Translated by A. F. Hort. New York: Loeb Classical Library/G. P. Putnam's Sons.

Thoreau, Henry David. 1862. "Wild Apples." *The Atlantic* 10 (5): 513–526.

Trifonov, V. A., N. I. Shishlina, Yu Lebedeva, J. van der Plicht, and S. A. Rishko. 2017. "Directly dated broomcorn millet from the northwestern Caucasus: Tracing the late Bronze Age route into the Russian steppe." *Journal of Archaeological Science: Reports* 12: 288–294.

Tuncle, G., M. J. R. Nout, and L. Brimer. 1993. "The effects of grinding, soaking and cooking on the degradation of amygdalin of bitter apricot seeds." *Food Chemistry* 53: 447–451.

Vainshtein, S. 1980. *Nomads of South Siberia: The Pastoral Economies of Tuva.* Translated by M. Colenso. Cambridge: Cambridge University Press.

Valamoti, S. M. 2016. "Millet, the later comer: On the tracks of *Panicum miliaceum* in prehistoric Greece." *Archaeological and Anthropological Sciences* 8 (1): 51–63.

Valamoti, S. M., and G. Jones. 2010. "Bronze and oil: A possible link between the introduction of tin and lallemantia to northern Greece." *Annual of the British School at Athens* 105: 83–96.

Van der Veen, Marijke. 2011. *Consumption, Trade and Innovation: Exploring the Botanical Remains from the Roman and Islamic Ports at Quseir al-Qadim, Egypt.* Frankfurt: Africa Magna Verlag.

Van Driem, George. 1999. "Neolithic correlates of ancient Tibeto-Burman migrations." In *Archaeology and Language II*, edited by Roger Blench and M. Spriggs, 67–102. London: Routledge.

Van Driem, George. 2002. "Tibeto-Burman phylogeny and prehistory: Languages, material culture and genes." In *Examining the Farming/Language Dispersal Hypothesis*, edited by Peter Bellwood and Colin Renfrew, 233–249. Cambridge: McDonald Institute for Archaeology.

Vaughan, D. A., B. Lu, and N. Tomooka. 2008. "The evolving story of rice evolution." *Plant Science* 174 (4): 394–408.

Ventresca Miller, A., E. Usmanova, V. Logvin, S. Kalieva, I. Shevnina, and A. Logvin. 2014. "Subsistence and social change in central Eurasia: Stable isotope analysis of populations spanning the Bronze Age transition." *Journal of Archaeological Science* 42: 525–538.

Victor, Mair, ed. 2012. *The "Silk Roads" in Time and Space: Migrations, Motifs and Materials*. Philadelphia: University of Pennsylvania Press.

Vishnu-Mittre. 1972. "Neolithic plant economy at Chirand, Bihar." *Palaeobotanist* 22 (1): 18–22.

Vogel, Hans Ulrich. 2013. *Marco Polo Was in China: New Evidence from Currencies, Salts and Revenues*. Boston: Brill.

von Bothmer, R., T. van Hintum, H. Knupfer, and K. Sato. 2003. *Diversity in Barley: Developments in Plant Genetics and Breeding 7*. Amsterdam: Elsevier.

Wagner, M., W. Xinhua, T. Pavel, A. Ailijiang, C. Bronk Ramsey, M. Schultz, T. Schmidt-Schultz, and J. Gresky. 2011. "Radiocarbon-dated archaeological record of early first millennium B.C. mounted pastoralists in the Kunlun Mountains, China." *Proceedings of the National Academy of Sciences* 108: 15733–15738.

Walsh, Judith E. 2006. *A Brief History of India*. New York: Infobase Publishing.

Wang, B. 1983. "Excavations and preliminary research on the remains from Gumugou on the Kongque River." *Social Science in Xinjiang* 1: 117–127. (In Chinese.)

Wang, Ch'ung. 1907. *Lun-Heng*. Translated by Alfred Forke. Shanghai: Kelly and Walsh.

Wang, Jiajing, Liu Li, Terry Ball, Linjie Yu, Yuanqing Li, and Fulai Xing. 2016. "Revealing a 5,000-y-old beer recipe in China." *Proceedings of the National Academy of Sciences* 113 (23): 6444–6448.

Wang, Z.-H., and E.-J. Zhuang. 2001. *China Fruit Monograph: Peach Flora*. Beijing: China Forestry Press.

Watkins, Ray. 1976. "Cherry, plum, peach, apricot, and almond: *Prunus* spp. (Rosaceae)." In *Evolution of Crop Plants*, edited by N. W. Simmonds, 242–247. New York: Longman.

Watson, Andrew M. 1983. *Agricultural Innovation in the Early Islamic World*. Cambridge: Cambridge University Press.

Watt, James C. Y., Jiayao An, Angela F. Howard, Boris I. Marshak, Bai Su, and Feng Zhao. 2004. *China: Dawn of a Golden Age, 200–750 A.D.* New York: Metropolitan Museum of Art.

Waugh, Daniel C. 2007. "Richthofen's 'Silk Roads': Toward the archaeology of a concept." *The Silk Road* 5 (1): 1–10.

Weber, Steve A. 1991. *Plants and Harappan Subsistence: An Example of Stability and Change from Rojdi*. Boulder, CO: Westview.

Weiss, E., and D. Zohary. 2011. "The Neolithic Southwest Asian founder crops: Their biology and archaeobotany." *Current Anthropology* 52 (supplement 4): S237–S254.

Wertmann, Patrick. 2015. *Sogdians in China: Archaeological and Art Historical Analyses of Tombs and Texts from the 3rd to the 10th Century AD*. Berlin: Philipp von Zabern GmbH.

Willcox, G. 1991. "Carbonized plant remains from Shortughai, Afganistan." In *New Light on Early Farming: Recent Developments in Palaeoethnobotany*, edited by J. M. Renfrew, 139–153. Edinburgh: Edinburgh University Press.

Willcox, G. 2013. "The roots of cultivation in southwestern Asia." *Science* 341: 39–40.

Willcox, George. 1994. "L'archéobotanique de Miri Qalat: Makran (The archaeology of Miri Qalat)." Unpublished report.

Wood, Frances. 1998. *Did Marco Polo Go to China?* Boulder, CO: Westview.

Wood, Frances. 2002. *The Silk Road: Two Thousand Years in the Heart of Asia*. Berkeley: University of California Press.

Wu, X., N. F. Miller, and P. Crabtree. 2015. "Agro-pastoral strategies at Kyzyltepa, an Achaemenid site in southern Uzbekistan." *Iran* 53: 93–117.

Wu, Zhengyi, Peter Ravens, and Missouri Botanical Gardens. 2006. *Flora of China*. St. Louis: Missouri Botanical Gardens.

Xenophon. 1922 [ca. 365 BC]. *Anabasis: Xenophon in Seven Volumes*. Translated by Carleton L. Brownson. Cambridge, MA: Harvard University Press.

Xu, T. 1982. "Origin and evolution of cultivated barley in China." *Acta Genetica Sinica* 9: 440–446.

Xue, Y. 2010. "A Preliminary Investigation on the archaeobotanical material from the site of Haimenkou in Jianchuan County." PhD diss., Peking University.

Yakubov, Yo. 1979. *Pargar in the VII–VIII Centuries AD: The Upper Zerafshan in the Early Middle Ages*. Dushanbe: A. Donesha Institute of the Academy of Science of the Tajik SSR. (In Russian.)

Yang, C. C., Y. Kawahara, H. Mizuno, J. Wu, T. Matsumoto, and T. Itoh. 2012. "Independent domestication of Asian rice followed by gene flow from japonica to indica." *Molecular Biology and Evolution* 29: 1471–1479.

Yang, Fuquan. 2005. "The 'Ancient Tea Horse Caravan Road,' the 'Silk Road' of southwest China." *The Silk Road* 2 (1).

Yang, S., Y. Wei, P. Qi, and Y. Zheng. 2008. "Sequence polymorphisms and phylogenetic relationships of Hina gene in wild barley from Tibet, China." *Agricultural Sciences in China* 7 (7): 796–803.

Yang, X., C. Liu, J. Zhang, and H. Lü. 2009. "Plant crop remains from the outer burial pit of the Han Yangling Mausoleum and their significance to early Western Han agriculture." *Chinese Scientific Bulletin* 54 (10): 1738–1743.

Yang, Xiaoyan, Zhiwei Wan, Linda Perry, Houyuan Lu, Qiang Wang, Chaohong Zhao, Jun Li, et al. 2012. "Early millet use in northern China." *Proceedings of the National Academy of Sciences* 109: 3276–3730.

Zeven, A. C., and J. M. de Wet. 1982. *Dictionary of Cultivated Plants and Their Regions of Diversity*. Wageningen: Centre for Agricultural Publishing and Documentation.

Zhao, Z. 2009. "Eastward spread of wheat into China: New data and new issues." In *Chinese Archaeology: Volume Nine*, edited by Q. Liu and Y. Bai, 1–9. Beijing: China Social Press.

Zhao, Z. 2011. "New archaeobotanic data for the study of the origins of agriculture in China." *Current Anthropology* 52: S295–S304.

Zheng, Yunfei, Gary W. Crawford, and Xugao Chen. 2014. "Archaeological evidence for peach *(Prunus persica)* cultivation and domestication in China." *PLOS One* 9 (9): e106595.

Zheng, Yunfei, Guoping Sun, Ling Qin, Chunhai Li, Xianhong Wu, and Xugao Chen. 2010. "Rice fields and modes of rice cultivation between 5000 and 2500 BC in east China." *Journal of Archaeological Science* 36: 2609–2616.

Zheng, Yunfei, Leping Jiang, and Jianming Zheng. 2004. "Study on the remains of ancient rice from Kuahuqiao in Zhejiang Province." *Chinese Journal of Rice Science* 18 (2): 119–124. (In Chinese.)

Zohary, D. 1999. "Monophyletic vs. polyphyletic origin of crops found in the Near East." *Genetic Resources in Crop Evolution* 46: 133–142.

Zohary, D., M. Hopf, and E. Weiss. 2012. *Domestication of Plants in the Old World: The Origin and Spread of Domesticated Plants in Southwest Asia, Europe, and the Mediterranean Basin*. 4th ed. Oxford: Oxford University Press.

INDEX